T0295313

FOCUS ON
ARTHROPODS RESEARCH

ANIMAL SCIENCE, ISSUES AND RESEARCH

Additional books and e-books in this series can be found on Nova's website under the Series tab.

ANIMAL SCIENCE, ISSUES AND RESEARCH

FOCUS ON ARTHROPODS RESEARCH

MIRKO MESSANA
EDITOR

nova
science publishers
New York

NOTICE TO THE READER

Library of Congress Cataloging-in-Publication Data

ISBN: 978-1-53614-343-0

Published by Nova Science Publishers, Inc. † New York

CONTENTS

PREFACE

Plant-insect interaction has remained stable for more than 300 million years, a fact that correlates with these groups' capacity to connect with their counterparts for reproduction, protection and feeding, among other things. *Focus on Arthropods Research* opens with an analysis of the various factors that influence these communities. Some studies suggest that genetic, chemical and morphological variability of host plants are the most influential factors.

The following chapter examines whether it is possible to infer the types and durations of embryonized larval stages in higher decapod species based on equivalent free-living stages of dendrobranchiate prawns, assuming the dendrobranchiate development pattern is ancestral to that of the entire Decapoda. Stage-specific development times of dendrobranchiate crustaceans were obtained from previous studies and analyzed across species to quantify the proportions of total development spent in each stage and phase.

Some arthropods, like the beetle *Scaurus punctatus* (Coleoptera: Tenebrionidae) are considered beneficial in terrestrial ecosystems of southern Europe due to their detritivore activity. There is little information available on their parasites and their effect on the beetles. Thus, the closing chapter provides new insight on aspects such as epizootiology, biology and the impact of gregarines in these insects.

Chapter 1 - Plant-insect interaction has maintained stable for more than 300 million years, a fact that has been related with these groups capacity to escape or associate with their counterparts for reproduction, protection and feeding, among others. Arthropod communities are influenced by various factors; however, some studies suggest that genetic, chemical and morphological variability of host plants are the factors which influence the most on arthropod community structure. In contrast, intrinsic insect characteristics, such as the exoskeleton, wings, and feeding preferences can explain their evolutionary success. In general, it has been proposed that proximal (ecological) and distal (evolutionary) factors are responsible for population abundance and distribution and for the community structure and functioning. Among proximal factors, the authors can mention: diversity (genetic and specific), interactions, disturbances, geographical and seasonal variations and edaphic factors. Among distal causes the authors can point out: natural selection, coevolution and adaptive radiation. For all the aforementioned, in this chapter the authors will describe and discuss factors the influence plant-insect interactions which are very important to elucidate the reasons for the ecological and evolutionary success of both groups.

Chapter 2 - Many species of decapod crustaceans that are important to fisheries or ecological communities belong to the suborder Pleocyemata, the so-called "higher" Decapoda that includes crabs, lobsters, and shrimps. There is considerable incentive to study the embryonic and larval development of these arthropods because they are important to recruitment. However, within this suborder only stages in the latter two of the four ancestral larval phases of decapod development are free-living (mysis and decapodid), while all stages in the earlier two phases (nauplius and protozoea) have become "embryonized" and occur within the egg. Due to stage-specific variations in development rates, this makes understanding and predicting development and hatch times of such decapods very difficult, especially since the precise number and characteristics of embryonized stages are unknown. However, within the decapod suborder Dendrobranchiata, the so-called "lower" Decapoda that includes prawns in the superfamilies Penaeoidea and Sergestoidea, all larval stages in all four phases are free-living. This chapter examines whether it is possible to infer the types and durations of embryonized larval stages in higher decapod species based on equivalent free-living stages of dendrobranchiate prawns, assuming the dendrobranchiate development pattern is ancestral to that of the entire Decapoda. Stage-specific development times of dendrobranchiate crustaceans were obtained from previous studies and analyzed across species to quantify the proportions of total development spent in each stage and phase. Then, egg and larval development times of selected higher decapods were also collected from previous studies. Whether the proportion of total development spent by each of these species in their free-living larval stages relative to the total duration of all embryonized stages matched the equivalent proportion determined from the dendrobranchiate species examined was then tested. If these proportions matched, then dendrobranchiate-derived proportions could be used to estimate the durations of embryonized stages within higher decapods. This provides information that could be validated by later studies, for example by conducting egg dissections at predicted time periods to confirm the presence and types of different stages. If confirmed, this technique may allow for improved predictions of egg development rates and hatch times of commercially important higher decapods.

Chapter 3 - Some arthropods, like the beetle *Scaurus punctatus* (Coleoptera: Tenebrionidae) are considered beneficial in terrestrial ecosystems of southern Europe, due to their detritivore activity. There is little information available on their parasites and the effect they cause on beetles. This chapter provides a new insight on aspects such as epizootiology, biology and impact of gregarines in these insects.

Epizootiological studies in this laboratory have led to the first finding of *Gregarina wharmani* and *Gregarina* sp (n°2) in *S. punctatus* beetles from Alcalá de Henares (Madrid, Spain). Since these organisms were described first in Israel and Maroc, this report represents the first finding in Europe. Interestingly, fully developed gamonts of *G. wharmani* show a sturdy pellicle or epicyte (a superficial layer) that stands the passage through the insect's intestinal tract. Therefore, this gregarine stage can be diagnosed by microscopical examination of feces.

A previous study conducted in the authors' laboratory has described in detail the morphology of *G. ormierei*. As a continuation of such investigations, some new aspects of the biology and morphology of *G. ormierei* are presented here:

1. Sporozoite release from the oocyst relies on a mechanism related to that seen in some Ciliates.
2. In the small phenotype, a cytoplasmic tail anchors the gamonts in syzygy to the insect's peritrhropic membrane,
3. Undehyscent gametocysts of both phenotypes are many times mature and contain viable oocysts capable of infecting insects,
4. Trophozoites show first a globular epimerite that later becomes conical/tubular. Such a new evidence suggests that *G. ormierei* and *G. cavalierina* might be the same organism. Histological examination of beetle gut tissue is neccesary to confirm such hypothesis.

Diverse environmental factors, like temperature and humidity, affect the populations of *G. ormierei*. Gregarine prevalence in beetles reaches a peak in late spring and decreases in summer. The life cycle duration is 10-25 days, depending on gregarine phenotype. Finally, to monitor possible detrimental effects of *G. ormeirei* on beetles, protozoan load and different morpho-biological characteristics of naturally infected and uninfected insects were analysed for correlation. There was significant negative correlation ($P < 0.05$) between gregarine load and beetle weight, body length and abdomen width, whereas positive correlation existed with both bold behavior and amputations ($P < 0.05$). The present results suggest that *G. ormierei* causes harmful effects in *S. punctatus*. Such finding is ecologically meaningful, since this abundant insect is involved in the natural transformation of decaying organic material in soil.

In: Focus on Arthropods Research   ISBN: 978-1-53614-343-0
Editor: Mirko Messana   © 2018 Nova Science Publishers, Inc.

Chapter 1

PROXIMAL AND EVOLUTIONARY FACTORS THAT INFLUENCE ARTHROPOD COMMUNITY STRUCTURE ASSOCIATED TO VASCULAR PLANTS

Efraín Tovar-Sánchez[1,], Elgar Castillo-Mendoza[1,2], Leticia Valencia-Cuevas[1], Miriam Serrano-Muñoz[1] and Patricia Mussali-Galante[3]*

[1]Laboratorio de Marcadores Moleculares, Centro de Investigación en Biodiversidad y Conservación, Universidad Autónoma del Estado de Morelos, Cuernavaca, Morelos, México
[2]Posgrado en Ciencias Biológicas, Universidad Nacional Autónoma de México, México City, México
[3]Laboratorio de Investigaciones Ambientales, Centro de Investigación en Biotecnología, Universidad Autónoma del Estado de Morelos, Cuernavaca, Morelos, México

* Corresponding Autor Email: efrain_tovar@uaem.mx.

ABSTRACT

Plant-insect interaction has maintained stable for more than 300 million years, a fact that has been related with these groups capacity to escape or associate with their counterparts for reproduction, protection and feeding, among others. Arthropod communities are influenced by various factors; however, some studies suggest that genetic, chemical and morphological variability of host plants are the factors which influence the most on arthropod community structure. In contrast, intrinsic insect characteristics, such as the exoskeleton, wings, and feeding preferences can explain their evolutionary success. In general, it has been proposed that proximal (ecological) and distal (evolutionary) factors are responsible for population abundance and distribution and for the community structure and functioning. Among proximal factors, we can mention: diversity (genetic and specific), interactions, disturbances, geographical and seasonal variations and edaphic factors. Among distal causes we can point out: natural selection, coevolution and adaptive radiation. For all the aforementioned, in this chapter we will describe and discuss factors the influence plant-insect interactions which are very important to elucidate the reasons for the ecological and evolutionary success of both groups.

Keywords: arthropod communities, distal factors, proximal factors

1. INTRODUCTION

Studying the interactions between living beings and their environment has interested human beings since they appeared on earth, even though the formal study of ecology started less than 200 years ago (Egerton, 2001). Recently, ecological studies acquired great importance due to the increasing degradation of ecosystems all around the world (Brehm et al., 2005), a development that has altered the interactions between different biological groups and endangers the dynamics that sustain most ecosystems (Díaz and Cabido, 2001).

One of the main challenges of studying ecology is to recognize and understand the mechanisms that enable species to coexist with each other, and to understand how their interactions affect the structure and functioning of biological communities. One of the approaches that have been used to understand these mechanisms is the grouping of species by ecological similarities (Sheley and James, 2010). Lindeman (1942) proposes the classification of species based on their position in the trophic chain: producers, consumers and decomposers. This classification is directly based on the interaction between animals and the plants from which they obtain resources. This chapter will use the definition of functional groups of De Bello et al. (2010), who grouped organisms according to the characteristics that are similar between them and that lead them to respond in a similar way to changes in the environment and/or to have a similar impact on ecosystem processes. Numerous studies support this type of classification (e.g., Hochwender and Fritz, 2004; Franks et al., 2009; Morais and Cianciaruso, 2014). This definition of functional group allows researchers to contextualize, in a broader way, their understanding of how the presence of different species of arthropods and their intra and interspecific interactions influence the structure and functioning of arthropod communities.

This chapter describes how proximate (ecological) and distal (evolutionary) causation factors have influenced the distribution and abundance of arthropod populations and the structure and functioning of arthropod communities. The proximate causes that will be analyzed include plant diversity, associational susceptibility, plant productivity, plant genetic diversity, chemical composition, structural complexity, environmental gradients, seasonality and disturbances. The distal causes include coevolution, hybridization and adaptive radiation. The chapter will focus on examples of arthropod species associated with oak species and will describe the main factors that influence the interaction between plants and insects and how this relationship contributes and/or has contributed to the ecological and evolutionary success of both groups.

2. BIOTIC FACTORS

2.1. Plant Diversity and Associational Susceptibility

Arthropods maintain a close relationship with their host plants, from which they obtain a wide range of benefits (e.g., food, shelter from predators or adverse conditions, places for sexual display, etc.) (Strong et al. 1984; Schoonoven et al., 2005). It has been proposed that plant diversity may be important in determining the diversity of the associated fauna (Hunter and Price, 1992; Siemman et al., 1998; Scherber et al., 2010). The argument is that a greater diversity of plants provides a broader range of resources and conditions that can sustain a greater number of associated species (Siemann et al., 1998; Haddad et al., 2001; Crutsinger et al., 2006; Vehviläinen et al., 2008). Field and experimental studies have confirmed this hypothesis by reporting an increase in the diversity of phytophagous arthropods as a result of the increase in diversity of the associated plants (Siemman, 1998; Knops et al., 1999; Hawkins and Porter, 2003). This increase in the spectrum of resources and conditions associated with plant diversity not only reduces the competition between arthropod species with similar requirements but can also facilitate the arrival and colonization by new species that exploit other available resources and conditions (Schowalter et al., 2011), which results in greater diversification.

Moreover, it has been reported that the diversity of plant species can have a positive influence on the richness of parasitoid arthropod species and their predators (Hunter and Price 1992; Siemman et al., 1998; Wojtowicz et al., 2014; Valencia-Cuevas et al., 2017). For example, Brown and Ewel (1987) proposed the phenomenon known as "associational susceptibility," which suggests that host plants spatially associated with heterospecific neighbors can sustain a community of herbivorous arthropods with greater abundance and diversity. This phenomenon is expected to occur when generalist arthropods benefit from the wide range of resources and conditions provided by diverse plant communities (Unsicker et al., 2008). It would also occur when the focal plant is the least preferred host but grows near the preferred host plant (Atsatt and O'Dowd, 1976); this promotes the mobility of arthropod species to neighboring plants after using the preferred host plant (White and Whitham, 2000).

The greater diversity of herbivores associated with host plants that coexist in floristically complex plant communities has been reported in temperate Mexican forests with respect to communities of ectophagous arthropods associated with the canopy of *Quercus crassifolia* (Tovar-Sánchez et al., 2015a), *Quercus crassipes* and *Q. rugosa* (Tovar-Sánchez et al., 2015b), and with respect to the community of gall-inducing endophagous insects (Cynipidae) associated with the canopy of *Q. castanea* through a gradient of species richness of red oak (section: *Lobatae*) in temperate forests of the center of Mexico (Valencia-Cuevas et al., 2017).

This suggests that a diverse community of plant species favors a greater diversity of the associated arthropod communities by making available for them a broader range of resources and conditions. This information should be considered when planning the management and conservation of diverse ecosystems.

2.2. Plant Productivity: Biomass

All organisms need energy to synthesize the molecules required to perform survival, growth and reproduction processes. The ability to obtain energy is an essential factor for individuals that is associated with their level of adaptation (Schowalter, 2011). In terrestrial ecosystems, plants are the main responsible for transforming solar energy into chemical energy through photosynthesis. Part of this energy is stored in plants in the form of organic matter or biomass; the rate at which this biomass is produced is known as primary productivity. Vegetable biomass can be used for energy by heterotrophic organisms, including arthropods. Thus, a higher primary productivity leads to a greater availability of resources for consumer species, increasing their abundance and the number of species associated with plants (Srivastava and Lawton, 1998).

Herbivore arthropods obtain the matter and energy they need to carry out their vital functions directly from plants. Predatory or parasitoid arthropods benefit indirectly from plant biomass because they use herbivores as a resource and can even respond directly to changes in vegetation (Siemman, 1998; Begon et al., 2006; Haddad et al., 2009). This dynamic should be considered when trying to understand the close relationship between plants and arthropods from different trophic levels, as well as the influence of plant biomass on the functioning and structure of arthropod communities (Strong et al., 1984).

In general, the biomass of plant communities can vary spatially and temporally (Begon et al., 2006), which has been explained as a result of the heterogeneity of environmental conditions and biotic processes (Valencia-Cuevas and Tovar-Sánchez, 2015). In turn, the characteristics of arthropod communities such as the abundance, richness and diversity of species respond to variations in plant biomass, that is, variations in conditions and in the availability of resources (Crutsinger et al., 2006; Haddad et al., 2009; Tomas et al., 2011; McArt et al., 2012). An example of the influence of plant biomass on arthropod communities can be seen in temperate ecosystems, where the phenology of plants is influenced by precipitation patterns. Plant biomass increases (production of branches, leaves and fruits) during the rainy season, which also creates the necessary conditions for the growth of various species of epiphytic plants in the canopy (Valencia-Cuevas and Tovar-Sánchez, 2015). Arthropod communities associated with the oak canopy (Quercus) of temperate forests have shown an increase in species abundance, richness and diversity when the range of resources and conditions becomes broader as a result of the increase in the biomass of host plants hosts during the rainy season (Forkner et al., 2004; Southwood et al., 2005; Tovar-Sánchez, 2009; Tovar-Sánchez et al., 2013; Tovar-Sánchez, 2015a).

At a spatial level, differences in biotic or abiotic factors may create differences in the amount of biomass generated by plant communities (Valencia-Cuevas and Tovar-Sánchez, 2015). Changes in biomass and plant productivity have been reported as a result of the changes in temperature and precipitation that occur along altitudinal gradients (Sundqvist et al., 2013). This has consequences for the abundance, diversity and richness of arthropod species (Körner, 2007; Lessar et al., 2011; Sundqvist et al., 2013; Bernardou et al., 2015).

Biological processes can also affect the production of plant biomass and, in turn, the arthropod communities associated with plants. Some studies have found that the diversity of plant species (Tilman et al., 1996; Cardinale et al., 2007; Haddad et al., 2009) and genotypes (Crutsinger et al., 2006; 2008 a, b) favors plant productivity [due to niche complementarity or facilitation (Hooper et al., 2005)], which benefits species richness and abundance of arthropods (Johnson and Agrawal, 2005; Johnson et al., 2006; Crutsinger et al., 2006; 2008a, b).

2.3. Plant Genetic Diversity

Genetic diversity is defined as the magnitude of genetic variability at the individual, population or species level (Nason, 2002). It is considered the raw material for evolution through natural selection (Fisher, 1930) and a fundamental source of biodiversity (Huges et al., 2008). In the last 20 years, different studies have found evidence that the genetic diversity of host plants is an important ecological factor that can influence the structure of the animal communities associated with them (Whitham et al., 2012; Crustinger et al., 2016). This evidence is associated with the recognition that the phenotypic characteristics of plant populations present substantial genetic diversity (Geber and Griffen, 2003). High-impact genes with influence at the community level have already been identified using QTLs (quantitative trait loci); they include genes that are responsible for the phenological pulses involved in the formation of buds (Frewen et al., 2000), tree growth and architecture, (Bradshaw and Stettler, 1995), resistance to pathogens (Newcombe and Bradshaw, 1996), and production of secondary metabolites (Shepherd et al., 1999; Freeman et al., 2008).

The study of plant-arthropod interactions has made it possible to demonstrate the influence of genes at the community level. The preferred study systems include founding species and their associated phytophages (herbivores) (Whitham et al., 2003, 2006, 2012). Founding species are plant species that structure communities by creating locally stable conditions, providing resources for other species and contributing to the modulation and stabilization of ecosystem processes (Ellison et al., 2005). In general, the genetic diversity of host plants is positively and significantly related to the diversity, richness and relative abundance of the associated herbivore communities. An example of this pattern can be found in poplars (Wimp et al., 2004; Bangert et al., 2005, 2006, 2008; Compson et al., 2016), oaks (Tovar-Sánchez and Oyama, 2006b; Tovar-Sánchez et al., 2013, 2015a, b; Valencia-Cuevas et al., 2017), eucalyptus (Dungey et al., 2000), and willows (Hochwender and Fritz, 2004). Even in poplars and oaks, the response of phytophagous organisms to the genetic diversity of host plants has been consistent regardless of the host species, type of forest and geographic scale (population and region). It has been suggested that an increase in the genetic diversity of host plants can generate changes in their morphological (Lambert et al., 1995; González-Rodríguez et al., 2004; Tovar-Sánchez and Oyama, 2004), phenological (Hunter et al., 1997), architectural (Martinsen and Whitham, 1994; Whitham et al., 1999; Bangert et al., 2005), and chemical (Fritz, 1999) characteristics, broadening the range of resources and conditions that can be exploited by herbivores.

Likewise, it has been observed that the genetic identity of host plants is an important regulator of the structure of arthropod communities, since genetically similar hosts sustain associated communities of genetically similar arthropods (Bangert et al., 2006; Compson et al., 2016)This fact has been explained by considering that genetically more similar populations have greater similarity in their physical, chemical and phenological characteristics, which will favor the establishment of more similar arthropod communities (Bangert and Whitham, 2007). The ability of herbivorous arthropods to discriminate between plant genotypes has been observed in different plant species, including poplars (Wimp et al., 2005; Compson et al., 2016), eucalyptus (Dungey et al., 2000), willows (Hochwender and Fritz, 2004) and oaks (Tovar-Sánchez and Oyama, 2006b).

The aforementioned studies have analyzed communities of specialized (endophagus) and generalist (ectophagous) herbivores, showing that the first group is more sensitive to the effects of the genetic diversity of the host plants (Bangert et al., 2005, 2006, 2008; Shuster et al., 2006; Tovar-Sánchez and Oyama, 2006b; Valencia-Cuevas et al., 2017). Endophagous herbivores are an important functional group that inhabits the canopy of trees, and includes insects such as gall-formers, leaf miners and leaf-rollers, which are characterized by living within the tissues of leaves and feeding on the mesophyll tissue (Cornell, 1990). It has been proposed that the heritable phenological and chemical signals that are produced by host plants determine the selection of oviposition and gall formation sites by endophagous herbivores (Abrahamson et al., 1993). For example, when choosing oak trees, gall-forming wasps of the Cynipidae family choose specific species, organs and tissues (Stone, 2002). This high degree of specialization and the close relationship of these insects with the host plant could be the reason why this group is so sensitive to the genetic diversity of the host (Tovar-Sánchez and Oyama, 2006b).

The genetic diversity of host plants not only has a direct effect on the associated herbivore communities; its influence can extend indirectly to the following trophic levels, creating a cascading effect through the ecosystem (Whitham et al., 2003, 2012). For example, an increase in the genetic diversity of the host plant may promote an increase in its architectural complexity and nutritional quality (Bailey et al., 2004; Glynn et al., 2004), which will favor an increase in the diversity, abundance and quality of the associated herbivores (Bailey et al., 2006; Valencia-Cuevas et al., 2017) which, in turn, will increase the intensity of predation and the degree of parasitism (Sarfraz et al., 2008).

In the coming years, a major challenge will be to understand the connections between evolutionary and ecological processes, given the continuing loss of genetic diversity throughout the world (Butchart et al., 2010) and the potential consequences on biological communities of the changes in the patterns of genetic variation caused by large selective events (Genung et al., 2011). Recognizing the influence of the genetic diversity of host plants on ecological processes constitutes a valuable contribution to ecological theory; the study of arthropods and their host plants has undoubtedly made a significant contribution to this theoretical advance.

2.4. Chemical Composition: Nutritional Quality and Secondary Metabolites

The plant chemicals that can influence the associated arthropod communities can be divided in two categories: food and defense (Strong, 1984). An example of food chemicals is foliar nitrogen, a critical component for phytophagous insects (Strong et al., 1984); its concentration is positively and significantly associated with the growth, reproduction and survival rate of herbivorous insects (Mattson Jr., 1980). The content of nitrogen in the leaves can vary due to leaf ontogeny or between plant species (Jeffries et al., 2006), altering the feeding preferences of herbivorous insects (Coley and Barone, 1996; Marquis and Lill, 2010). For example, low nitrogen content has been associated with a low preference and reduced performance of herbivorous insects, as the palatability of a plant depends on the carbon/nitrogen ratio of the leaves (Schädler et al., 2003).

From a community-focused perspective, there are several studies that show the importance of the nitrogen content of host plants on their associated insect communities. Some studies have reported a positive and significant relationship between the concentration of foliar nitrogen and the density of herbivorous insects (e.g., leaf miners, chewing insects, gall-formers and leaf rollers) in various species of oak: *Quercus alba* (Wold and Marquis, 1997), *Q. prinus*, *Q. rubra* (Forkner and Hunter, 2000), *Q. dentata* (Nakamura et al., 2008), *Q. germinata*, *Q. laevis* (Cornelissen and Stiling, 2006, 2008), *Q. alba*, *Q. coccinea* and *Q. velutina* (Marquis and Lill, 2010). Likewise, a greater richness of species of chewing insects (Lepidoptera) has been reported when there is a greater amount of foliar nitrogen in *Q. crispula* (Murakami et al., 2005, 2007, 2008).

It has also been well documented that plants produce other chemical substances (e.g., oxalic acid, alkaloids, phenolic compounds, toxic lipids, flavonoids, tannins and lignins) that act as defenses or insect attractants (Becerra et al., 2001). In response to these signals or chemical defenses, there can be changes in the structure of arthropod communities associated with the canopy of trees (Inoue et al., 2003).

Tannins stand out among secondary metabolites for their defensive role and their effect on herbivorous insects and on the structure of their communities (Feeny, 1970). It has been reported that tannins reduce the growth and survival of phytophagous insects (Kause et al., 1999; Lill and Marquis, 2001), produce lethal deformities (Barbenhenn and Martin, 1994) and increase parasitism rates (Faeth and Bultman, 1986). At the community level, a negative relationship has been reported between the concentration of tannins in host plants and the abundance and richness of herbivorous insects (e.g., chewing insects, leaf miners and gall-formers) that inhabit the canopy of the following oak species: *Q. alba*, *Q. velutina* (Le Corff and Marquis, 1999; Forkner et al., 2004), *Q. crispula* (Murakami et al., 2005, 2007, 2008), *Q. germinata*, *Q. laevis* (Cornelissen and Stiling, 2006, 2008) and *Q. gambelii* × *Q. grisea* (Yarnes et al., 2008).

There are reports that the species of individual plants has a significant effect on the concentration of nitrogen and secondary metabolites (Suomela and Ayres, 1994), which suggests that species variability affects the foraging activity and spatial distribution of arthropods. The concentration of nitrogen and secondary metabolites may depend on the following factors: 1) the genotype of the host plant (Glynn et al., 2004), 2) the environmental conditions (Larsson et al., 1986), and 3) the resources of the host plant (Ricklefs, 2008).

Gall-formers (Cynipidae) are a group of insects that is sensitive to the differences in leaf chemistry between species of oaks. Abrahamson et al. (1998, 2003) found that the structure of gall-wasp communities was different and particular in each of six different species of oak (*Q. laevis*, *Q. myrtifolia*, *Q. inopina*, *Q. chapmanii*, *Q. geminata* and *Q. minima*). Similar results were reported for the complex *Q. crassipes* × *Q. crassifolia* in Mexico (Tovar-Sánchez and Oyama, 2006b), for *Q. infectoria* and *Q. brantii* (Nazemi et al., 2008) in Iran, and for *Q. castanea* and *Q. crassipes* in Mexico (Tovar-Sánchez et al., 2013). Researchers suggest that this sensitivity is explained by the close relationship between plants and insects, which is behind the high degree of specialization of these insects with respect to the chemicals of their host oaks.

Moreover, oaks have been observed to show seasonal variations in foliar chemistry. Some examples have been documented in *Q. robur* (Feeny, 1970; Salminen et al., 2004), *Q. alba*, *Q. velutina* (Le Corff and Marquis, 1999), *Q. alba* (Lill and Marquis, 2001), *Q. crispula* (Murakami et al., 2005, 2007, 2008), *Q. germinata* and *Q. laevis* (Cornelissen and Stiling, 2006, 2008). These studies found temporal variations in the nutritional quality of the leaves; as the ontogenetic development of leaves progressed, the content of tannins and lignins increased and the content of water and nitrogen decreased (Feeny, 1970). Several studies have shown that the richness, diversity, abundance and biomass of the arthropods associated with the canopy of oak trees decrease as the season progresses, while the structure of the communities changes in response to variations in the chemistry of the host plants (e.g., Forkner et al., 2004; Southwood et al., 2004; Yarnes et al., 2008).

Studying the chemistry of host plants and of its influence on plant-insect interactions is crucial for understanding the structure of arthropod communities and the preferences of individual arthropods regarding oviposition site, feeding habits, ontogenetic performance and change of host.

2.5. Structural Complexity

Since plant communities determine the physical structure of different environments, they have a great influence on the structure of the associated animal communities (Strong et al., 1984; Halaj et al., 2000; Tews et al., 2004). It has been suggested that structurally more complex environments offer a wider range of available habitats and shelters and create the conditions for the occurrence of speciation events as a result of the adaptation of species to various environmental conditions (Halaj et al., 2000; Tews et al., 2004; Kallimanis et al., 2008; Antonelli and Sanmartın, 2011), all of which promotes the coexistence, persistence and diversification of species (Stein et al., 2014). In the case of arthropods, the abundance and architecture of the plants with which they are associated (e.g., shape and size of leaves, shoots, branches and epiphytic plants, as well as the texture of the stem or bark; Halaj et al., 2000; Sobek et al., 2009) constitute a very important element of the structure and complexity of their habitat. Its importance lies in the fact that the heterogeneity of the habitat is related to the availability of basic resources for herbivores such as: food, shelter and foraging, oviposition and sexual display sites (Halaj et al., 2000; Novotny et al., 2006)For example, some studies have shown that the presence of epiphytic plants (e.g., orchids, bromeliads, ferns, mosses, lichens, etc.), which differ substantially between them in their structure, growth habit and function, increases the structural complexity of the tree canopy, offering a great diversity of microhabitats and resources (Ishii et al., 2004) that can be used by the arthropods associated with the trees.

Another factor that increases the complexity of the habitat of arthropods associated with plants is the richness of species in plant communities. Plant communities rich in species have also greater richness, diversity and abundance of arthropod species (Sobek et al., 2009). The positive relationship between the richness of plant species and the richness of arthropod species has been widely documented (Gaston, 1991; Siemman, 1998; Knops et al., 1999; Hawkins and Porter, 2003; Vehviläinen et al., 2008). An increase in plant diversity represents an increase in the diversity of resources available for herbivores, which allows more consumer species to coexist (Hutchinson, 1959), since it is more likely that a particular resource is available to a particular consumer. Under this scenario, the diversity of herbivores is promoted by plant diversity (Chown et al., 1998; Novotny et al., 2006; Kumar et al., 2009). Moreover, plant diversity can indirectly influence predator communities through its effect on the diversity, abundance and quality of their prey (herbivores) (Chown et al., 1998; Scherber et al., 2010; Valencia-Cuevas et al., 2017).

Another scenario in which the effect of habitat complexity becomes evident is of the succession of plant communities. For example, the structural complexity of forests increases with their age, that is, mature forests are structurally more complex than young forests or plantations (Schowalter, 1995; Hardiman et al., 2011), since the former tend to have a greater number of large in terms of height and biomass, more tree species and different strata of vegetation (herbaceous, shrubs, trees) (Bazzaz, 1975;

Fernandes et al., 2010). Mature forests also contain trees of different ages, which gives them greater structural complexity (Ishii et al., 2004). In short, the presence of tree species with diverse physiognomy and growth patterns, as well as individuals of different sizes, generates horizontal and vertical variation, which promotes microenvironmental heterogeneity (Stein et al., 2014) and provides more diverse resources and conditions that can be exploited by arthropods.

An example of the positive effect of the structural complexity that age gives to a forest habitat on the richness of the arthropod fauna associated with the tree canopy was observed in the community of chewing insects associated with the canopy of *Q. alba* and *Q. velutina* (Marquis and Le Corff, 1997) and the community of lepidoptera associated with the canopy of *Quercus* spp. (Summerville and Crist, 2002, 2003). Furthermore, Marquis et al. (2002) showed that the abundance of shelter-building caterpillars (Lepidoptera) is related to the architecture of their host oaks (*Q. alba*). Their results showed a positive and significant relationship between the structural complexity of the canopy of this tree species, measured as the proportion of overlapping leaves, and the abundance of caterpillars, which indicates the importance of tree architecture for these herbivores.

The results of the different studies included here suggest that habitat complexity is a crucial factor favoring the diversity of arthropod species, which means that management and conservation plans aimed at the preservation of these organisms should contemplate the inclusion of this factor when designing strategies for the maintenance of biodiversity.

2.6. Interactions

Galls are considered a "hot spot" because, throughout the different stages of their development, they are the place where several species interact. Gall-inducing insects (Cynipini) are found there during the formation stage (Pujade-Villar, 2013). Galls are induced by the chemical action caused by the secretions and excretions of insect larvae, which control the development of galls; if the larvae die the development of galls stops (Folliot, 1977; Pujade-Villar, 2013).

Synergini insects also appear in galls, as inquilines or commensals. The synergini lost their ability to induce their own galls (Pénzes et al., 2012), but they can still induce their own larval chambers and create their own nutritional tissue; furthermore, they can kill inducer organisms by competing with them for space and/or food (Ronquist, 1999; Maldonado-López, et al., 2013). Chalcidoid insects (Chalcidoidea) can also be found during the formation of galls. They belong to six families (Eulophidae, Eupelmidae, Eurytomidae, Ormyridae, Pteromalidae and Torymidae), and their ecological function in galls is not yet completely understood, since they can act as phytophagous insects, predators, parasitoids and hyperparasitoids. It is thought that chalcidoids regulate the populations of gall-inducing insects, since they can kill between 40 and 100% of inducer insects (Gibson, 2006).

There are few studies on the primary fauna of galls; however, it is already possible to associate some genera of synergini and chalcidoids to certain gall-inducing insects. Examples of this have been documented by Serrano-Muñoz (2016), who mentions that the insects found in the galls induced by Trigonaspis oscura in *Q. rugosa* include the inquiline *Synergus* sp. and the chalcidoids *Baryscapus*, *Brasema*, *Eurytoma*, *Sycophila* and *Ormyrus*. Furthermore, the inquiline *Synergus* and the chalcidoids *Galeopsomyia*, *Eurytoma*, *Ormyrus* and *Sycophila* have been found in galls induced by *Atrusca pictor* in *Q. frutex*, while the inquiline *Synerus* and the chalcidoids *Baryscapus*, *Eurytoma*, *Galeopsomyia*, *Ormyrus* and *Torymus* were found in galls induced by *Disholcaspis potosina* in *Q. obtusata*. Once a gall is established, secondary fauna can be found in them, constituted by small arthropods that use the gall as a refuge and/or food (Pujade-Villar, 2013). Valencia-Cuevas et al. (2017) collected Diptera (Cecidomyiidae and Chloropidae), Lepidoptera (Bedellinae and Gelechiidae) and Hymenoptera (Apidea, Bethylidae, Braconidae, Figitidae and Sphecidae) from galls of *Q. castanea*. That was the first study carried out in Mexico that mentioned the secondary fauna that can be found in plant galls, providing a background for future studies. After a gall falls to the ground, it is consumed by soil organisms such as fungi and small arthropods. However, sometimes the galls are consumed by birds or mammals. It is expected that future studies expand our knowledge of the biological interactions centered around galls, and of the ecological function of each of their inhabitants.

3. ABIOTIC FACTORS

3.1. Environmental Gradients

The ability of arthropods to establish themselves in a given habitat is limited in part by variations in the physical environment (Menke and Holway, 2006). Factors such as temperature, humidity, precipitation, solar radiation and wind speed directly affect the survival, reproductive capacity and longevity of arthropods (Willmer, 1982). For example, the desiccation of insects in immature stages increases in environments with high temperatures and lower humidity (Willmer, 1982). In fact, it has been reported that insect larvae survive and grow better under conditions of higher humidity (Hunter and Willmer, 1989; Larsson et al., 1997). Likewise, it has been observed that herbivorous arthropods can be directly affected by solar radiation (light and heat), but also indirectly by affecting the nutritional quality of their host plants (Fukui, 2001). Finally, radiation and wind cause changes in the temperature of the air, which affects the survival and growth of arthropods (Porter and Gates, 1969). These changes in the physical environment can modify the behavior, physiology and morphology of individual arthropods, inducing alterations in the distribution and abundance of populations (McKinney, 2008) and, consequently, in the structure of arthropod communities (Valencia-Cuevas and Tovar-Sánchez, 2015).

Abiotic factors can also affect plant communities along environmental gradients, inducing changes in species richness, genetic diversity, abundance and total biomass (Begon et al., 2006); this creates heterogeneous habitats (in terms of conditions and resources), which in turn induce changes in the associated arthropod communities (Valencia-Cuevas and Tovar-Sánchez, 2015). If the composition and diversity of plant communities vary predictably across habitats and biogeographical zones (Gurevitch et al., 2002), it is reasonable to assume that the strength of plant-herbivore interactions can also vary. Two of the most recognized and studied environmental gradients are those associated with latitudinal and altitudinal changes (Gaston, 2000; Lomolino et al., 2006b). In general, climatic variables (temperature, humidity, precipitation), solar radiation and edaphic parameters (pH, availability of nutrients), among others, vary with latitude and altitude (Körner, 2007). Several studies have reported the response of various biological groups, including arthropods and their host plants, to environmental gradients (Hodkinson, 2005; Sundqvist et al., 2013; Bernadou et al., 2015). Changes in the abiotic factors described above affect the phenology, morphology, physiology and chemistry of host plants (Hodkinson, 2005), altering their ability to defend against herbivorous arthropods. (Pellisier et al., 2012). Thus, changes in altitude (Rodríguez-Castañeda et al., 2010; Beck et al., 2011; Sundqvist et al., 2013; Bernadou et al., 2015) and latitude (Gaston and Lawton 1988; Hillebrand 2004; Dyer et al., 2007; Kraft et al. 2011) correlate with changes in the diversity, composition and abundance of plant-associated arthropods. One of the most frequently observed patterns along altitudinal gradients is the decrease in species richness as altitude increases (Rabhek, 2005; Grytnes and McCain, 2007; McCain et al., 2011). However, some studies have found peaks of species richness at intermediate altitudes (Rahbek, 2005; Kessler et al., 2011). Both patterns have been documented in arthropod communities (Olson, 1994; Sanders et al., 2003).

A study carried out in Mexico along an altitudinal gradient in a temperate forest (*Abies-Quercus*) evidenced a decrease in the abundance and diversity of species of the arthropod community associated with the mulch forest, as well as changes in the composition of species along the gradient (Rodríguez-Domínguez, 2010). The author suggests that this response from the arthropod community can be partly explained by changes in factors such as temperature and in edaphic attributes that vary with altitude.

The study of the influence of abiotic factors along environmental gradients has served to understand the limits imposed by environmental conditions on the distribution of arthropods, which, in turn, helps to understand their biology and abundance (Andrew and Hughes, 2005; Sundqvist et al., 2013). Furthermore, the response of arthropod species and communities to changes in abiotic factors associated with environmental gradients can be used as a predictive tool to understand the potential impact of climate change (Hodkinson, 2005) and the future of arthropod biodiversity (Fukami and Wardle, 2005; McCain and Colwell, 2011).

3.2. Seasonality

The abundance and distribution of arthropods can change across space and time. An arthropod species will only be present in a given habitat when it has the capacity to reach it (dispersion), when the resources and conditions necessary for its establishment become available and when its competitors, predators or parasitoids allow it (Begon et al., 2006). Thus, the temporal sequence of the appearance and disappearance of an arthropod species in a given habitat depends on how the influence of resources, conditions and enemies changes over time. Needless to say, the factors that affect vegetation will have an effect on arthropod communities (Valencia-Cuevas and Tovar-Sánchez, 2015).

Highly synchronized phenological events are characteristic of deciduous temperate forests, where the foliage of most tree species regrows in spring and falls in the autumn (Strong et al., 1984). These seasonal cyclical changes in plant species change the distribution and abundance of the arthropod species associated with the tree canopy. Different studies have documented the seasonal changes in the composition of arthropod communities and the decrease in the relative abundance and richness of arthropod species associated with the canopy of temperate forests (Gering et al., 2003; Tovar-Sánchez, 2009; Tovar-Sánchez et al., 2013). These responses have been explained by considering that the nutritional quality of the leaf tissue found in temperate forests decreases as the season progresses (Feeny, 1970); leaves become harder, their content of water and nitrogen content decreases and the concentration of tannins and fiber increases (Feeny, 1970). These changes in foliage characteristics contribute to the appearance of arthropod species with different feeding preferences (e.g., leaf-chewing insects at the beginning of the season, sucking insects at the end of the season; Strong et al., 1984; Southwood et al., 2004, 2005), changing the composition of arthropod communities.

Another characteristic of temperate forests is the seasonal pattern of the rainy season, which lasts six to seven months and is followed by a dry season that can last from five to six months (Rzedowski, 1978). These annual variations in the rainfall pattern affect the phenology of vegetation and its associated arthropod communities. The formation of branches, foliage and fruits, and the growth of epiphytic plants, increases during the rainy season, broadening the range of resources and conditions that can be exploited by the arthropods inhabiting the canopy. These events can create microclimatic changes (Basset and Novotny, 1999) and increase the heterogeneity of the habitats used by arthropods (Yarnes and Boecklen, 2005). Moreover, young leaves, which are less hard, of higher nutritional quality and with a smaller amount of chemical defenses, are more abundant during the rainy season (Kursar and Coley, 2003; Forkner and Marquis, 2004). Finally, the increase in plant biomass during the rainy season in temperate forests can create for foliage arthropods to colonize new trees (Basset et al., 1992).

Oaks are one of the most representative genera of temperate forests. Studies of the arthropod communities associated with the canopy of this plant group have documented the effect of their phenological changes on the communities associated with them. For example, as the seasons progress, a decrease in density, richness, diversity and biomass has been reported in the community of leaf-chewing insects associated with the canopy of *Q. alba* and *Q. velutina* in Missouri (Forkner et al., 2004), the community of herbivorous insects that inhabit the canopy of *Q. cerris*, *Q. ilex*, *Q. petraea* and *Q. robur* in France (Southwood et al., 2004, 2005), and the community of beetles associated with the canopy of *Quercus* spp. in Turkey (Şen and Gök, 2009).

Likewise, greater diversity, species richness, density and biomass have been reported during the rainy season in the collembola community living in Tillandsia spp., which grows associated with the canopy of *Quercus* spp. in a temperate forest in central Mexico (Palacios-Vargas and Castaño-Meneses, 2003). These changes have also been observed in the community of ectophagous insects associated with the canopy of *Q. laurina* and *Q. rugosa* (Tovar-Sánchez, 2009), and the canopy of *Q. castanea* and *Q. crassipes* (Tovar-Sánchez et al., 2013) in a temperate forest in central Mexico.

In general, the high sensitivity of arthropods to changes in biotic and abiotic parameters suggests its usefulness as bioindicators; they can be an important tool to understand and predict the effect of environmental change on biodiversity.

3.3. Disturbances

Disturbances are discrete events that alter the structure of populations, communities and ecosystems by changing the availability of resources and the prevailing conditions (White and Pickett, 1985). Succession is the process of recovery after a disturbance (Harper, 1977) in which biological communities experience changes in their composition, complexity, diversity and habits (Currano et al., 2011). In plant communities, disturbances act as promoters of succession, giving rise to vegetation mosaics with different degrees of structural complexity (White and Pickett, 1985; Siemann et al., 1998; Fernandes et al., 2010) and inducing the turnover of species. In sum, they affect the dynamics of biological communities (Hughes et al., 2007; Ilg et al., 2008; Gerisch et al., 2012). In the case of arthropods, they are affected, directly and indirectly, by the intensity, frequency, duration, and area of the disturbances (Currano et al., 2011). Directly when frequent disturbances maintain the diversity of biological communities at low levels by causing local extinction events and limiting the dispersion of species (Hanski, 1994). Indirectly by causing changes in the structure of plant communities, which affects the spatial and temporal patterns of the diversity of the arthropods associated with them (Fagan et al., 1999; Jeffries et al., 2006).

In general, biological communities are subject to the effects of disturbances of natural [e.g., fire, storms, hurricanes, floods, etc. (Dziock et al., 2006)] and anthropogenic [e.g., deforestation, agriculture, urbanization, etc. (Hirao et al., 2007)] origin. This section addresses only the effects of anthropogenic disturbances, the frequency and intensity of which have increased dramatically in recent years.

Several studies have documented that anthropogenic disturbances affect arthropod communities in several ways and with different intensity (Hill et al., 1995; Floren and Linsenmair, 2001; Currano et al., 2011); there is no unique pattern. Disturbance events can have negative or positive effects, or simply have no effect, on the structure of arthropod communities (Mackey and Currie, 2001). This range of responses can be explained by differences in the habitat requirements, dispersal abilities and distribution patterns of different arthropod species (Cooke and Roland, 2000; Gibb et al., 2013), as well as by differences in the scale and degree of the disturbances (Lewis, 2001; Niemelä et al., 2002), historical factors and heterogeneity of the sites (Bruno et al., 2003; Hamer et al., 2003). For example, the abundance and diversity of arthropod species that depend on resources that are only available after a disturbance (e.g., dead wood) could increase rapidly after the event, compared to those species that depend on microhabitats that are not regularly affected by disturbances (Gibb et al., 2013). In consequence, undisturbed sites can make biological communities more stable but less diverse.

The canopy of oaks and the arthropods associated with it constitute a useful system to illustrate the diversity of responses to disturbances of arthropod communities. A study by Tovar-Sánchez et al. (2003) compared three forests with different degrees of disturbance in the Valley of Mexico and found significant differences in the abundance and diversity of ectophagous arthropods associated with the canopy of *Q. castanea*, *Q. crassipes*, *Q. crassifolia*, *Q. greggii*, *Q. laeta* and *Q. rugosa*. Similar responses were reported regarding the richness of herbivorous insects associated with the canopy of *Q. alba* and the richness and abundance of leaf-chewing insects associated with the canopy of *Q. alba* and *Q. velutina* in Missouri (Forkner et al., 2006, 2008). Moreover, Summerville and Crist (2002, 2003) found changes in the composition of species, and a decrease in the richness thereof, in the lepidopteran community associated with the canopy of *Quercus* ssp. in recently felled forests compared to non-felled ones.

Other studies have reported that the arthropod fauna associated with the canopy of oaks does not respond to disturbances. For example, the composition and species richness of beetles associated with the canopy of *Quercus* spp. in forest fragments in Bulgaria with different degrees of urbanization (rural/suburban/urban) did not present differences between them (Niemelä et al., 2002). The authors suggest that local factors such as temperature, humidity or edaphic conditions may have played a more important role in the structure of the beetle community, considering that the degree of habitat disturbance that occurred in the three groups of forest fragments was moderate.

Some studies have reported that arthropods benefit from disturbances (Chust et al., 2007; Maldonado-López et al., 2015). Maldonado-López et al. (2015) reported greater abundance and diversity of gall-inducing insects of the Cynipidae family as the fragmentation of oak forests increased. The authors explain that these results can be explained by the dispersal capacity of these arthropods and by the increase in the quality of the host plants, measured in terms of the amount of leaves, buds and petioles, which are the sites where cynipids induce the formation of galls.

The future of biodiversity depends to a large extent on the generation of knowledge that is useful for managing ecosystems that have been altered by human activities. One of the most important things is to identify the species or groups of species that are sensitive to these alterations (Gardner, 2010). The sensitivity of arthropods to changes in their habitat is a very valuable attribute that makes this animal group a potential ecological indicator that can provide information on the conservation state of different ecosystems (Valencia-Cuevas and Tovar-Sánchez, 2015).

4. EVOLUTIONARY FACTORS

4.1. Coevolution

Since the publication of the study by Ehrlich and Raven (1964) on the coevolution between plants and animals, the scientific community became greatly interested in the study of ecological interactions, their possible evolutionary histories and their role in the structuring of biological communities (Oyama, 1999). The study by Ehrlich and Raven proposed the concept of coevolution as a mechanism that could explain the joint evolution of butterflies of the superfamily Papilionoidea and the plant species with which they were associated. They used this concept to explain the process by which plants develop new chemical defenses and insects evolve resistance or tolerance to these new defenses. The authors suggested that this type of reciprocal responses between plants and insects could be considered one of the most important regulating factors of diversity in terrestrial communities, one in which a crucial role is played by the biochemical innovation that takes place in plants (Ehrlich and Raven, 1964; Becerra et al., 2009; Futuyma and Agrawal, 2009; Thompson, 2013). To study this mechanism, Berenbaum (1983) analyzed the chemical composition of plants of the family Umbelliferae and the herbivores associated with plant species with different chemical composition. A characteristic of this group of plants is that they contain a chemical group known as coumarins, which have at least four derivative groups that differ in a single chemical radical. This study showed that the degree of toxicity of each derived chemical group increases with its degree of complexity, which suggests that this progressive increase in toxicity corresponds to a biosynthetic advance against herbivorous insects. Another contribution of the study carried out by Berenbaum was to show that the plant taxa with the most advanced coumarins were richer in species compared to those with less advanced chemicals. It also showed that the insect taxa that fed on plants with more complex coumarins were richer in species than other taxa that fed on plants with simpler coumarins. These results provided the first evidence of the important association between the evolution of plant-insect interactions and diversification events. Furthermore, the results obtained by this study inspired decades of work that aimed to document this type of evolutionary scenarios (Berenbaum, 2001).

Coevolution is a process that defines and redefines the interactions between different species (Thompson, 2001). Gall-inducing insects and their host plants constitute a good model to illustrate coevolution events. It has been suggested that gall formation emerged as a defense mechanism in plants to isolate potentially harmful insects and restrict their development (Ananthakrishan, 1984; Stone et al., 2002). However, insects evolved in response to this strategy, inducing changes in plant growth and living within its tissues (Ananthakrishnan, 1984). They managed to use galls to obtain food, protection against predators, drying and shelter for reproduction

(Fernandes and Price, 1988). This has led some researchers to suggest that the formation of galls is an evolutionary step towards a stronger relationship between genetic changes and the exploitation of plant species (Ananthakrishnan, 1984).

An example of this type of interaction can be observed in cynipid wasps (Hymenoptera: Cynipidae), which form galls in oak trees (Fagaceae: *Quercus*). The interaction between cynipid wasps and oaks has a long history of at least 30 million years (starting in the Oligocene or Miocene; Kinsey, 1930). An important characteristic of this group of insects is its specificity regarding the genus of the host plant and even the organs that are attacked by them, so that a certain species of cynipidae is associated only with a certain species or related group of plant species and induces galls constantly and exclusively in a single organ of the plant (Stone et al., 2002). Any organ and part of the plant can be attacked by arthropods, including roots, stems, buds, leaves, flowers and fruits (Ronquist and Liljeblad, 2001; Stone et al., 2002). Tannins are part of the defensive chemistry oaks. These chemical compounds stand out for their role as herbivore repellents (Feeny, 1970) and for their influence on the structuring of phytophagous insect communities (Inoue et al., 2003). Tannins have been reported to affect insect growth and survival (Kause et al., 1999), reducing their biomass (Lill and Marquis, 2001), producing lethal deformities (Barbenhenn and Martin, 1994) and increasing parasitism rates (Faeth and Bultman, 1986). However, this chemical barrier has not been effective in controlling the growth and development of gall-inducing cynipids. Considering that the generation time of oaks is much longer than that of gall wasps, it is not surprising that the latter have evolved different strategies to overcome the tannin barrier. Some insect species have developed powerful phenoloxidase systems, which are enzymes that have the ability to oxidize tannins (Nierenstein, 1930). Beside tannins, the defense mechanisms of oaks against the attack of cynipids include seasonal changes in their nutritional quality and in the turgidity of their leaves (Strong et al., 1984; Forkner et al., 2004). However, these insects have managed to evade these defenses by manipulating the levels of tannins and nutrients in the gall tissues (Fay and Hartnett, 1991; Bagatto et al., 1996; Hartley, 1998, Schönrogge et al., 2000). This scenario shows the way in which these two groups of organisms have coevolved over millions of years (Nieves-Aldrey, 1987).

Another model that can be used to identify reciprocal adaptive responses are the interactions between parasites and hosts (Beranbaum, 2001). The communities associated with the galls induced by cynipids (Hymenoptera, Cynipidae) belong to several trophic levels, forming complex networks composed of inquiline, parasitoid and successor insects (Askew, 1984). Galls constitute extended phenotypes of the genes of the wasps that induce these structures in the tissue of their host oak (Dawkins, 1982; Stone and Cook, 1998; Stone and Schönrogge, 2003). These structures have evolved into increasingly complex morphologies, the purpose of which, in part, has been to exclude the natural predators of these wasps, mainly parasitoid insects. It has been suggested that the relationship between parasitoid insects and their host wasps has been maintained through coevolution processes that have given rise to diverse communities that include a third of all animal species (Bailey et al., 2009). Parasitoid insects cause high mortality in gall wasps (Stone et al., 2002; Stone and Schönrogge, 2003); thus, natural selection could have favored adaptive responses by the host insects to reduce the rate of parasitism (Abrahamson and Weis, 1997; Stone and Schönrogge, 2003). Because the attack of parasitoids involves oviposition through the gall tissues, selection could have favored gall-forming insects with genes that induce the formation of structures in the galls that reduce or prevent such attacks (Stone and Schönrogge, 2003; Singer and Sireman, 2005; Abrahamson and Blair, 2008). The defensive phenotypes acquired by gall wasps have probably stimulated reciprocal evolutionary changes (Abrahamson and Weis, 1997; Agrawal, 2001) in the characteristics of parasitoid insects, such as the length of the ovipositor (Askew, 1965). Bailey et al. (2009) studied the effect of different host characteristics on 48 communities of parasitoids that attack gall-inducing wasps of the family Cynipidae in oak trees. The authors showed that gall attributes such as turgor, hairiness and the presence of sticky substances, as well as their position in the host plant, had a significant effect on the composition of the parasitoid insect community. It has been suggested that these changes in parasitoid insect communities reflect the action of different species; small parasitoid species attack during the early development of the galls, while larger ones with longer ovipositors attack during the later stages of gall development (Briggs and Latto, 1996; Abrahamson and Weis, 1997; Stone et al., 2002). These results support the hypothesis that the evolution of gall morphology has been an adaptive response of wasps to minimize the attacks from their associated parasitoids (Abrahamson and Weis, 1997; Stone and Schönrogge, 2003; Abrahamson and Blair, 2008). However, the results also suggest that parasitoids are still searching how to counteract these defenses (Wiebes-Rijks and Shorthouse, 1992; Stone et al., 2002).

4.2. Hybridization

Natural hybridization is a common phenomenon in plant species (Whitney et al., 2010) that has been recognized as a substantial evolutionary force that favors the process of species diversification and increases intraspecific genetic diversity (Rieseberg and Ellstrand, 1993; Whitham et al., 1999). In the last decades, hybrid zones have allowed to study the effect of the interspecific genetic flow on plant-insect interactions (Whitham, 1989; Dungey et al., 2000; Tovar-Sánchez and Oyama, 2006a, b; Yarnes et al., 2008; Valencia-Cuevas et al., 2017). Several studies have focused on the response of arthropods, particularly phytophagous ones, to the variations found in these areas (Boecklen and Spellenberg, 1990; Aguilar and Boecklen, 1992; Prezsler and Boecklen, 1994; Tovar-Sánchez and Oyama, 2006a, b; Yarnes et al., 2008). The host plants of arthropods have unique combinations of genetically based characteristics that could be associated with the oviposition preferences of the associated insects and the resistance characteristics of the plants (Boecklen and Spellenberg, 1990; Aguilar and Boecklen, 1992; Fritz, 1999). Thus, genetic variations that occur as a result of hybridization events may affect the distribution of herbivorous and pathogenic arthropods (Whitham et al., 1994; Fritz, 1999). For example, differences in the abundance or composition of arthropod communities in hybrid zones may be the result of the presence of susceptible hybrid genotypes, while in other areas the hybrids may be resistant, resulting in hybrid zones containing plants with greater or lesser resistance (Martinsen et al., 2000), which affects the structure of the arthropod community.

In general, phytophagous arthropods show four response patterns to the hybridization of their host plants: 1) Susceptibility: more insect species in hybrid hosts than in the parent species (Fritz et al., 1994; Whitham et al., 1994); 2) Dominance: hybrids sustain as many species of herbivores as some of the parent species (Fritz et al., 1994, 1996; Fritz, 1999); 3) Resistance: hybrids sustain less herbivores than the parent species (Boecklen and Spellenberg, 1990; Fritz et al., 1994, 1996; Fritz, 1999); 4) Additivity: hybrids sustain an intermediate number of insects compared to the parent species (Boecklen and Spellenberg, 1990; Fritz et al., 1994, 1996; Fritz, 1999). The presence, in phytophagous arthropods, of different response patterns to the hybridization of their host plants has been attributed to the extension, in time and space, of the geographical distribution of hybrid zones, environmental gradients, the genetic status of hybrids, morphological and chemical similarities between parent species, and the genetic mechanisms that determine the inheritance of resistance mechanisms in hybrids (Boecklen and Spellenberg, 1990; Strauss, 1994; Fritz, 1999; Whitham et al., 2003).

In the reviews carried out by Strauss (1994) and Whitham et al. (1999), the authors found that, in 152 cases analyzed, 79% of the taxa showed a significant response to the hybridization of the host plant, and the most frequent response was susceptibility of the hybrids, with 28% of the cases (43 studies). They also found that herbivores of both parent species accumulate in hybrids and hybrid zones.

Other factors that may affect the response of arthropods to the hybridization of their host plants are: 1) the level of genetic variation in the host plants present in hybrid zones, and 2) the pattern of introgression (Whitham et al., 1999). The highest genetic variation is expected to appear when all hybrid classes are present within a hybrid zone, so that any factor that eliminates one or more classes may have negative consequences on the levels of variation (Whitham et al., 1999). For example, it has been proposed that the greatest genetic diversity occurs as a result of bidirectional introgression, that is, when the first generation of hybrids (F1) are fertile, reproduce and form backcrosses with both parent species. A continuum of genotypes between both species would be expected as a result of the combinations and permutations of the genome of the two parental species (Tovar-Sánchez and Oyama, 2004). Nevertheless, it is also possible that fertile F1 hybrids cross with only one of the parent species (Keim et al., 1989), a process known as unidirectional introgression. Under this hybridization scenario, the continuum of hybrid genotypes exists only between the hybrids and one of the parent species, while a morphological and genetic vacuum is left between the hybrids and the other parent species, which results in a reduction of the genetic diversity of the hybrid zone. In the last scenario, F1 hybrids are sterile, survive by cloning, and genetic diversity is at its lowest level (Whitham et al., 1999). An example that illustrates the influence of introgression and genetic diversity in hybrid zones on arthropod communities is the study by Tovar-Sánchez and Oyama (2006b) in seven hybrid zones of the hybrid complex *Q. crassipes* × *Q. crassifolia* in Mexico. Their results showed that in the hybrid zone where bidirectional introgression was detected, the hybrids sustained the greatest richness of endophagous insects (Hymenoptera and Lepidoptera), compared to the hybrids in the other six hybrid zones where introgression was unidirectional. The authors also reported a positive and significant relationship between the genetic diversity of the hybrid zone and the diversity of endophagous arthropods associated with the same oak complex. They found the greatest genetic diversity in the hybrid zone where bidirectional introgression took place (Tovar-Sánchez and Oyama, 2004), which favors the species diversity of the community of endophagous insects. Whitham et al. (1999) proposed that the richness of arthropod species is highest in hybrid zones with bidirectional introgression, intermediate in hybrid zones formed by unidirectional introgression, and lowest in hybrid zones formed by sterile F1 hybrids, where no backcrossing with parent species occurs. This pattern has been explained by considering that hybrid genotypes are eliminated in the hybridization scenarios with unidirectional introgression and sterile hybrids, while phytophagous arthropods of the two parental species accumulate in the hybrid zones, affecting the diversity pattern of arthropod species (Whitham et al., 1999).

Hybridization is a mechanism that has allowed species of arthropods to change to a new species of host plant. It has been suggested that for this host change to occur, herbivores must be preadapted to switch to a new host species, but they do not because the new host is not present (Pre-adaptation hypothesis; Thomas et al., 1987). When herbivores are not pre-adapted to a new host plant, one or more key mutations must occur for phytophagous arthropods to recognize it as a new and better host (Mutation hypothesis; Jermy, 1984). Floate and Whitham (1993) proposed the hybrid bridge hypothesis that predicts that intermediate hybrid plants facilitate switching hosts from one species to another, since organisms associated with a particular plant can experiment and adapt gradually to the genome of another plant species. If a host species has an allopatric distribution with respect to another potential host, this creates a barrier that prevents phytophagous arthropods from switching hosts (Keim et al., 1989). The pre-adaptation hypothesis suggests that arthropods will not switch host species unless the hosts have a sympatric distribution (Thomas et al., 1987). Therefore, if two species hybridize, hybrid intermediaries become "spatial bridges" through which arthropods can switch to a new host plant species, even though the parent species have an allopatric distribution.

A study that supports the hypothesis of the hybrid bridge was carried out by Tovar-Sánchez and Oyama (2006b); they found that the hybrids that resulted from the cross of *Q. crassipes* and *Q. crassifolia* (*Q. × dysophylla*) hosted nine species of insects that usually inhabit the canopy of the two parental species.

Considering the frequency with which the phenomenon of hybridization occurs in plants, several studies suggest that hybrid zones are the centers of arthropod biodiversity (Whitham, 1989; Dungey et al., 2000; Tovar-Sánchez and Oyama, 2006a, b; Valencia-Cuevas et al., 2017). They also suggest that these areas are useful for exploring ecological and evolutionary processes at multiple levels (Strauss, 1994; Whitham et al., 1999). Conserving these areas becomes important because they have positive effects on the arthropod communities associated with species involved in hybridization events.

4.3. Adaptive Radiation

Arthropods represent more than half of the known species on earth (Hamilton et al., 2010); they have experienced several important ecological and evolutionary radiations (Condamine et al., 2016). Their evolution is associated with the great diversity of life forms and development strategies that can be found among them; this diversity has allowed them to occupy almost every ecological niche (Grimaldi and Engel, 2005). Their great diversity reflects the variety of adaptive transformations they have undergone under similarly varied environmental conditions (Schowalter, 2000). Thanks to the fossil record, it has been possible to determine that arthropods were able to survive the most severe mass extinction events, adapting to radical changes in terrestrial vegetation, continental rearrangements and changes in environmental parameters (Condamine et al., 2016). The most common hypotheses that have tried to explain the diversification of arthropods through their evolutionary history mention low rates of extinction and resilience to mass extinctions, as well as the acquisition of novel abilities that allowed them to occupy new niches (Labandeira et al., 1993; Grimaldi and Engel, 2005; Mayhew, 2007; Rainford et al., 2014).

It has been suggested that the diversity of morphological characters in the buccal apparatus, the appearance of wings, their small size, the presence of an exoskeleton and the process of complete metamorphosis constituted novelties that allowed insects to adapt to different environments and, consequently, to diversify (Strong et al., 1984; Labandeira et al., 1994; Condamine et al., 2016). An analysis of the fossil record of insects at the family level revealed that the appearance of wings is associated with a high rate of species origination, while the process of complete metamorphosis is associated with a higher rate of diversification (Labandeira et al., 1994). These results are consistent with those of a phylogenetic study involving 82% of all insect families that identified 45 changes in the diversification rate corresponding to the tree of life of these organisms. The authors mention that two of these changes are major ones and are related to the origin of flight and the emergence of the process of complete metamorphosis (Rainford et al., 2014). It has also been suggested that environmental and physical factors resulting from climatic or geological events, and even the emergence of new ecological niches, could have induced adaptive responses in arthropods (Condamine et al., 2016).

The case of herbivorous insects and their host plants can illustrate some of the scenarios mentioned above. Insects are the most dominant group on Earth in terms of species richness and ecological function (Wilson, 1992). Within this group, those with phytophagous habits have the highest species richness (Rivera, 1991; Schowalter, 2000); of these, the most important orders are Lepidoptera and Orthoptera, since close to 99% of their species are phytophagous (Strong et al., 1984). According to the fossil record, terrestrial plants and insects appeared in the Devonian period about 380-400 million years ago (Rohdendorf and Raznitsin, 1989; Wooton, 1981). Furthermore, it is believed that the periods of greatest increase in plant diversity are the Devonian period (mid Cretaceous), the Upper Cretaceous and the Tertiary (Knoll et al., 1979 These periods of diversification of plant species coincide with the increase in the diversity of phytophagous insects, probably as a response to the emergence of new niches as a result of increased plant diversity (Strong et al., 1984). Comparing the diversity of herbivorous insects with that of their non-herbivorous relatives suggest that the acquisition of a plant-based diet is associated with speciation and diversification (Mitter et al., 1988). Recent phylogenetic-molecular studies support the above hypothesis by indicating that orders such as Hymenoptera, Lepidoptera and Orthoptera (Hunt et al., 2007; Moreau et al., 2006; Ahrens et al., 2014) became widely diversified during the Cretaceous in response to newly formed niches (Mayhew, 2007). It has also been reported that an important part of the phytophagous beetle fauna emerged during radiation events caused by the appearance of new angiosperm lineages (Farrell, 1998). Furthermore, the appearance in the fossil record of gall-inducing wasps coincides with the origin of the main lineages of oaks that constitute their host plants (Ronquist and Liljeblad, 2001). Their diversification process was probably stimulated by the acquisition of novel traits that allowed them to exploit available resources (Strong et al., 1984; Cornell, 1989; Hespenheide, 1991; Labandeira et al., 1993; Grimaldi and Engel, 2005).

Moreover, it has been suggested that the Devonian and Carboniferous plant communities were structurally more complex than their predecessors (Sporne, 1975). Some researchers speculate that the increasing complexity of plants could have contributed to the increasing insect diversity (Strong et al., 1984). It has been suggested that the appearance of giant arborescent plants at the end of the Devonian period had two important consequences for the evolution of insects: a) it favored the evolution of winged insects and b) it stimulated the diversification of insects during the Carboniferous period (Strong et al., 1984). These hypotheses have been supported by contemporary studies that have shown that habitat complexity influences the diversity of insect communities (Halaj et al., 2000; Tews et al., 2004; Kallimanis et al., 2008; Antonelli and Sanmartin, 2011).

There are also studies that suggest that changes in global temperature and fluctuations in the content of atmospheric O_2 and CO_2 could be associated with diversification events of herbivorous insects (Labandeira, 2006). These possible associations could have been mediated by the effect of environmental factors on the vegetation that constituted the main resource for herbivorous insect communities.

Evolutionary novelties promoted by competition are considered one of the most important causes of adaptive radiation (Schluter, 2001). This may be due to the fact that the struggle for resources modifies the morphological and ethological characteristics that favor the use of unexploited resources (Abrams, 2000), resulting in genetic and phenotypic divergence, which would eventually cause species radiation. However, it has been suggested that interspecific competition has had limited influence on the diversification of arthropods (Strong et al., 1984). This hypothesis has been proposed with respect to insects. Information from the fossil record and contemporary insect communities suggest that interspecific competition has probably had little weight in the diversification of this group (Lawton and Strong, 1981; Strong et al., 1984; Denno et al., 1995), since their natural enemies and other factors keep their population low compared to the availability of resources, reducing the need for competition (Hairston et al., 1960). Several studies have reported that some species of herbivorous insects facilitate the presence of other species of insects by creating entry points, shelters or other modifications in the host plant (Waltz and Whitham, 1997; Martinsen et al., 2000; Lill and Marquis, 2003). However, Ronquist and Liljeblad (2001) suggest that inquiline insects associated with the galls induced by wasps of the family Cynipidae may have emerged as a result of a competitive interaction with the latter, losing the ability to induce galls and becoming obligatory inquilines (Ronquist, 1994).

The development of adaptations that allowed them to maintain feeding relationships with plants is another important factor in the diversification of phytophagous (Janz, 2011; Condamine et al., 2016). Plants are the most important food resource in terrestrial ecosystems (Begon et al., 2006). However, the consumption of plant biomass by herbivores represents a loss of energy for the plant. Under this scenario, plants seek to break their interaction with herbivorous organisms. At the same time, the herbivores that depend on certain plants seek to maintain their interaction with them by adapting to the changes of their host plants. In this process, herbivores have managed to diversify. In Europe, for example, the species *Quercus robur* has been reported to have 20 different types of galls formed by 20 different species of gall-forming wasps of the family Cynipidae (Crawley, 1997), possibly due to the existence of very specific defense mechanisms that have been suppressed in various forms by wasps, leading to radiation events. The insects associated with a single plant species represent an extreme case of adaptive radiation through the differential use of resources (Oyama, 2012). A study conducted by Abrahamson et al. (2003) reported that the community of gall-inducing wasps associated with six species of oak showed different assemblages in response to the presence of different chemical compounds in each host oak. This suggests that the presence of different defense chemicals in the host oaks may have contributed to the adaptive radiation of their associated herbivores.

CONCLUSION

When we talk about loss of species, we refer to those species that have already been taxonomically identified. According to recent estimates, arthropods are the most diverse biological group on the planet, but there are not many researchers to study them. This creates the urgent need to train human resources interested in the study, description and analysis of arthropods.

In general, studies aimed at the conservation of species are often focused on preserving plant species and "charismatic" animal species. Thus, arthropods are usually not given much attention.

There should be more studies focused on characterizing the structure of arthropod communities, not only because of the high levels of species richness and abundance they contain, but also because of the important ecological role they play (as functional groups). In natural conditions, all interactions (biotic, chemical, genetic and environmental) affect arthropods; however, their most important interaction is with host plants (whether they are specialists or generalists), since their survival, adaptation, evolution and diversity seem to be directly related to the range of resources and conditions that plants "offer" them directly (e.g., herbivores) or indirectly (e.g., parasites, parasitoids, decomposers, etc.). Further studies should aim towards the following objectives: first, to describe and make an inventory of the diversity of arthropods. Second, to study the factors that modify the structure of their communities. Third, to document the interactions between species of arthropods belonging to different trophic levels. Fourth, to describe the ecological role of arthropod species. These academic efforts should go hand-in-hand with efforts to conserve as many ecosystems as possible in order to preserve the greatest possible number of species. The data obtained from studies such as this one suggests that arthropods can be used as bioindicator species of environmental quality.

REFERENCES

Abrahamson, W. G. & Blair, C. P. (2008). Sequential radiation through host-race formation: herbivore diversity leads to diversity in natural enemies. In: K. Tilman, editor. *Specialization, speciation, and radiation: the evolutionary biology of herbivorous insects* (first edition, pp. 188–202). Berkeley, California: University of California Press.

Abrahamson, W. G., Brown, J. M., Roth, S. K., Sumerford, D. V., Horner, J. D., Hess, M. D., Torgerson, S., How, S., Craig, T. P., Packer, R. A. & Itami, J. (1993). Gallmaker speciation: an assessment of the roles of host-plant characters, phenology, gallmaker competition, and natural enemies. In: P. W. Price, W. J. Mattson & Y. N. Baranchikov (Eds.), *The ecology and evolution of gall forming insects* (first edition, pp. 208–222). St. Paul, MN: North Central Forest Experimental Station, Forest Service, USDA.

Abrahamson, W. G., Hunter, M. D., Melika, G. & Price, P. W. (2003). Cynipid gall-wasp communities correlate with oak chemistry. *Journal of Chemical Ecology, 29,* 209–223.

Abrahamson, W. G., Melika, M. D., Scrafford, R. & Csóka, P. (1998). Gall-inducing insects provide insights into plant systematic relationship. *American Journal of Botany, 85,* 1159–1165.

Abrahamson, W. G. & Weis, A. E. (1997). *Evolutionary ecology across three trophic levels: goldenrods, gallmakers and natural enemies.* Princeton, New Jersey: Princeton University Press.

Abrams, P. A. (2000). Character shifts of prey species that share predators. *The American Naturalist, 156,* 45–61.

Aguilar, J. M. & Boecklen, W. J. (1992). Patterns of herbivory in the *Quercus grisea* × *Quercus gambelii* species complex. *Oikos, 64,* 498–504.

Agrawal, A. A. (2001). Ecology - phenotypic plasticity in the interactions and evolution of species. *Science, 294,* 321–326.

Ahrens, D., Schwarzer, J. & Vogler, A. P. (2014). The evolution of scarab beetles tracks the sequential rise of angiosperms and mammals. *Proceedings of the Royal Society of London. Series B, 281,* 20141470.

Ananthakrishnan, T. N. (1984). *Biology of Gall Insects.* First edition. New Delhi, Delhi: Oxford & IBH.

Andrew, N. R. & Hughes, L. (2005). Arthropod community structure along a latitudinal gradient: implications for future impacts of climate change. *Australian Journal of Ecology, 30,* 281–297.

Antonelli, A. & Sanmartın, I. (2011). Why are there so many plant species in the Neotropics? *Taxon, 60,* 403–414.

Askew, R. R. (1965). The biology of the British species of the genus *Torymus* Dalman (Hymenoptera: Torymidae) associated with galls of Cynipidae (Hymenoptera) on oak, with special reference to alternation of forms. *Transactions of the Royal Entomological Society of London, 9,* 217–32.

Askew, R. R. (1984). The biology of gallwasps. In: T.N. Ananthakrishnan, editor. *The Biology of Galling Insects* (first edition, pp. 223–271). New Delhi, Delhi: Oxford & IBH.

Atsatt, P. R. & O' Dowd, D. J. (1976). Plant defense guilds. *Science,* 93: 24–29.

Bagatto, G., Paquette, L. C. & Shorthouse, J. D. (1996). In£uence of galls of Phanacis taraxaci on carbon partitioning within common dandelion, *Taraxacum officinale. Entomologia Experimentalis Et Applicata, 79,* 111-117.

Bailey, R., Schönrogge, K., Cook, J. M., Melika, G., Csóka, G., Thuróczy, C. & Stone, G. N. (2009). Host Niches and Defensive Extended Phenotypes Structure Parasitoid Wasp Communities. *PLoS Biology, 7(8),* e1000179.

Bailey, J. P., Schweitzer, J. A., Rehill, B. J., Lindroth, R. L., Martinsen, G. D. & Whitham, T. G. (2004). Beavers as molecular geneticists: a genetic basis to the foraging of an ecosystem engineer. *Ecology, 85,* 603–608.

Bailey, J. P., Wooley, S. C., Lindroth, R. L. & Whitham, T. G. (2006). Importance of species interactions to community heritability: a genetic basis to trophic level interactions. *Ecology Letters, 9,* 78–85.

Bangert, R. K., Allan, G. J., Turek, R. J., Wimp, G. M., Meneses, N., Martinsen, G. D., Keim, P. & Whitham, T. G. (2006). From genes to geography: a genetic similarity rule for arthropod community structure at multiple geographic scales. *Molecular Ecology, 15,* 4215–4228.

Bangert, J. K., Londsford, E. V., Wimp, G. M., Shuster, S. M., Fischer, D., Schweitzer, J. A., Allan, G. J., Bailey, J. K. & Whitham, T. G. (2008). Genetic structure of a foundation species: scaling community phenotypes from the individual to the region. *Heredity, 100,* 121–131.

Bangert, J. K., Turek, R. J., Martinsen, G. D., Wimp, G. M., Bailey, J. K. & Whitham, T. G. (2005). Benefits of conservation of plant genetic diversity on arthropod diversity. *Conservation Biology, 19,* 379–390.

Bangert, J. K. & Whitham, T. G. (2007). Genetic assembly rules and community phenotypes. *Evolutionary Ecology, 21,* 549–560.

Barbenhenn, R. V. & Martin, M. M. (1994). Tannin sensitivity in larvae of *Malacossomadisstira* (Lepidoptera): roles of the peritrophic envelope and midgut oxidation. *Journal of Chemical Ecology, 20,* 1985–2001.

Basset, Y., Aberlenc, H. P. & Delvare, G. (1992). Abundance and stratification of foliage arthropods in a lowland rainforest of Cameroon. *Ecological Entomology, 17,* 310–318.

Basset, Y. & Novotny, V. (1999). Species richness of insect herbivore communities on *Ficus* in Papua New Guinea. *Biological Journal of the Linnean Society, 67,* 477–499.

Bazzaz, F.A. (1975). Plant Species Diversity in Old-Field Successional Ecosystems in Southern Illinois. *Ecology, 56,* 485-488.

Becerra, J. X., Noge, K. & Venable, D. L. (2009). Macroevolutionary chemical escalation in an ancient plant-herbivore arms race. *Proceedings of the Royal Society of London. Series B, 106,* 18062–18066.

Becerra, J. X., Venable, D. L., Evans, P. H. & Bowers, W. S. (2001). Interactions between chemical and mechanical defenses in the plant genus *Bursera* and their implications for herbivores. *American Zoologist, 41,* 865–876.

Beck, J., Brehm, G. & Fiedler, K. (2011). Links between the environment, abundance and diversity of Andean moths. *Biotropica, 43,* 208–217.

Begon, M., Towsend, C. R. & Harper, J. L. (2006). *Ecology: from individuals to ecosystems* (Fourth edition). Oxford, United Kingdom: Blackwell Publishing.

Berenbaum, M. (1983) Coumarins and caterpillars: A case for coevolution. *Evolution, 37,* 163–179.

Berenbaum, M. (2001). Plant-Herbivore Interactions. In: C. W. Fox, D. A. Roff & D. F. Fairbairn (Eds.), *Evolutionary ecology. Concepts and case*

38   *E. Tovar-Sánchez, E. Castillo-Mendoza, L. Valencia-Cuevas et al.*

studies (first edition, pp. 303–330). Albany, New York: Oxford University Press.

Bernadou, A., Espadaler, X., Le Goff, A. & Fourcassié, V. (2015). Ant community organization along elevational gradients in a temperate ecosystem. *Insectes Sociaux, 62,* 59–71.

Boeclken, W. J. & Spellenberg, R. (1990). Structure of herbivore communities in two oak (*Quercus* spp.) hybrid zones. *Oecologia, 85,* 92–100.

Bradshaw, H. D. Jr. & Stettler, R. F. (1995). Molecular genetics of growth and development in *Populus*. IV. Mapping QTLs with large effects on growth, form, and phenology traits in a forest tree. *Genetics, 139,* 963–973.

Brehm, G., Pitkin, L.M., Hilt, N. & Fiedler, K. (2005) Montane Andean rain forests are a global diversity hotspot of geometrid moths. *Journal of Biogeography, 32,* 1621–1627.

Briggs, C. J. & Latto, J. (1996). The window of vulnerability and its effects on relative parasitoid abundance. *Ecological Entomology, 21,* 128–140.

Brown, B. J. & Ewel, J. J. (1987). Herbivory in complex and simple tropical successional ecosystems. *Ecology, 68,* 108–116.

Bruno, J. F., Stachowicz, J. J. & Bertness, M. D. (2003). Inclusion of facilitation into ecological theory. *Trends in Ecology & Evolution, 18,* 119–125.

Butchard, S. H. M., Walpole, M., Collen, B., Van Strien, A., Almond, R. E. A., Jonathan E. M., Bastian Bomhard, B., Brown, C., Bruno, J., Carpenter, K. E., Carr, G. M. Chanson, J., Chenery, A. M., Csirke, J., Davidson, N. C., Dentener, F., Foster, M., Galli, A., Galloway, J. N., Genovesi, P., Gregory, R. D., Hockings, M., Kapos, V., Lamarque, J., Leverington, F., Loh, J., McGeoch, M. A., McRae, L., Minasyan, A., Hernández Morcillo, M., Oldfield, T. E. E. Daniel Pauly, Suhel Quader, Carmen Revenga, John R. Sauer, Benjamin Skolnik, Spear, D., Stanwell-Smith, D., Stuart, S. N., Symes, A., Tierney, M., Tyrrell, T. D., Vié, J., Watson, R. (2010). Global biodiversity: Indicators of recent declines. *Science, 328,* 1164–1168.

Cardinale, B. J., Wright, J. P., Cadotte, M. W., Carroll, I. T., Hector, A., Srivastava, D. S., Loreau, M. & Weis, J. J. (2007). Impacts of plant diversity on biomass production increase through time because of species complementarity. *Proceedings of the National Academy of Sciences of the United States of America, 104*, 18123–28.

Chown, S. L., Gremmen, N. J. M. & Gaston, K. J. (1998). Ecological biogeography of Southern Ocean islands: species-area relationships, human impacts, and conservation. *The American Naturalist, 152*, 562–575.

Chust, G., Garbin, L. & Pujade-Villar, J. (2007). Gall wasps and their parasitoids in cork oak fragmented forest. *Ecological Entomology, 32*, 82–91.

Coley, P. D. & Barone, J. A. (1996). Herbivory and plant defenses in tropical forest. *Annual Review of Ecology, Evolution, and Systematics, 27*, 305–335.

Compson, Z. G., Hungate, B. A., Whitham, T. G., Meneses, N. P., Busby, E. T. Wojtowicz, A., Ford, C., Adams, K. J. & Marks, J. C. (2016). Plant genotype influences aquatic- terrestrial ecosystem linkages through timing and composition of insect emergence. *Ecosphere, 7*, e01331.

Condamine, F. L., Clapham, M. E. & Kergoat, G. J. (2016). Global patterns of insect diversification: towards a reconciliation of fossil and molecular evidence? *Scientific Reports, 6*, 1–13.

Cooke, B. J. & Roland, J. (2000). Spatial analysis of large-scale patterns of forest ten caterpillar outbreaks. *Ecoscience, 7*, 410–422.

Cornelissen, T. & Stiling, P. (2006). Responses of different herbivore guilds to nutrient addition and natural enemy exclusion. *Ecoscience, 13*, 66–74.

Cornelissen, T. & Stiling, P. (2008). Cumpled distribution of oak leaf mines between and within plants. *Basic and Applied Ecology, 96*, 7–77.

Cornell, H. V. (1989). Endophage-ectophage ratios and plant defense. *Evolutionary Ecology, 3*, 64–76.

Cornell, H. V. (1990). Survivorship, life history, and concealment: a comparison of leaf miners and gall formers. *American Naturalist, 136*, 581–597.

Crawley, M. J. (1997). Plant-herbivore dynamics. In: M. J. Crawley, editor. *Plant Ecology* (Second edition, pp 401–474). Oxfordshire, Oxford: Blackwell Scientific Publications.

Crutsinger, G. M. (2016). A community genetics perspective: opportunities for the coming decade. *New Phytologist*, 210: 65–70.

Crutsinger, G. M., Collins, M. D., Fordyce, J. A., Gompert, Z., Nice, C. C. & Sanders, N. J. (2006). Plant genotypic diversity predicts community structure and governs an ecosystem process. *Science, 313*, 966–968.

Crutsinger, G. M., Collins, M. D., Fordyce, J. A. & Sanders, N. J. (2008a). Temporal dynamics in non-additive responses of arthropods to host-plant genotypic diversity. *Oikos, 117*, 255–264.

Crutsinger, G. M., Reynolds, W. N., Classen, A. T. & Sanders, N. J. (2008b). Disparate effects of plant genotypic diversity on foliage and litter arthropod communities. *Oecologia, 158*, 65–75.

Currano, E. D., Jacobs, B. F., Pan, A. D. & Tabor, N. J., (2011). Inferring ecological disturbance in the fossil record: a case study from the late Oligocene of Ethiopia. *Palaeogeography, Palaeoclimatology, Palaeoecology, 309*, 242–252.

Dawkins, R. (1982). *The extended phenotype: the gene as the unit of selection*. First edition. Oxfordshire, Oxford: Oxford University Press.

De Bello, F., Lavorel, S., Diaz, S., Harrington, R., Cornelissen J. H. C., Bardgett, R. D., Berg, M. P., Cipriotti, P., Feld, C. K., Hering, D., Martins da Silva, P., Potts, S.G., Sandin, L., Sousa, J. P., Storkey, J., Wardle, D.A. & Harrison, P. A. (2010). Towards an assessment of multipleecosystem processes and services via functional traits. *Biodiversity and Conservation, 19*, 2873–3893.

Denno, R. F., McClure, M. S. & Ott, J. R. (1995). Interspecific interactions in phytophagous insects: competition reexamined and resurrected. *Annual Review of Entomology, 40*, 297–331.

Diaz, S. & Cabido, M. (2001). Vive la difference: plant functional diversity matters to ecosystem processes. *Trends in Ecology and Evolution, 16*, 646–55.

Dungey, H. S., Potts, B. M., Whitham, T. G. & Li, H. F. (2000). Plant genetics affects arthropod community richness and composition:

evidence from a synthetic eucalypt hybrid population. *Evolution, 54,* 1938–1946.

Dyer, L. A., Singer, M. S., Lill, J. T., Stireman, J. O., Gentry, G. L., Marquis, R. J., Greeney, H. F., Wagner, D. L., Morais, H. C., Diniz, I. R., Kursar, T. A. & Coley, P. D. (2007). Host specificity of Lepidoptera in tropical and temperate forests. *Nature, 448,* 696–699.

Dziock, F., Henle, K., Follner, F. & Scholz, M. (2006). Biological indicators systems in floodplains—a review. *International Review of Hydrobiology, 191,* 292–313.

Ehrlich, P. R. & Raven, P. H. (1964). Butterflies and plants: a study in coevolution. *Evolution, 18,* 586-608.

Egerton, F. N. (2001). A History of the Ecological Sciences: Early Greek Origins. *Bulletin of the Ecological Society of America*, 89, 93–97.

Ellison, A., Bank, M. S., Clinton, B. D., Colburn, E. A., Elliott, K., Ford, C. R., Foster, D. R., Kloeppel, B. D., Knoepp, J. D., Lovett, G. M., Mohan, J., Orwig, D. A., Rodenhouse, N. L., Sobczak, W. V., Stinson, K. A., Stone, J. K., Swan, C. M., Thompson, J., Holle, B. V. & Webster, J. R. (2005). Loss of foundation species: consequences for the structure and dynamics of forested ecosystems. *Frontiers in Ecology and the Environment, 3,* 479–486.

Faeth, S. H. & Bultman, T. L. (1986). Interacting effects of increased tannin levels on leafmining insects. *Entomologia Experimentalis et Applicata, 40,* 297–300.

Fagan, W. F., Cantrell, R. S. & Cosner, C. (1999). How habitat edges change interactions. *The American Naturalist, 153,* 165–182.

Farrell, B. D. (1998). 'Inordinate fondness' explained: why are there so many beetles? *Science, 281,* 555–559.

Fay, P. A. and Hartnett, D. C. (1991). Constraints on growth and allocation patterns of *Silphium integrifolium* (Asteraceae) caused by a cynipid gall wasp. *Oecologia, 88,* 243–250.

Feeny, P. (1970). Seasonal changes in oak leaf tannins and nutrients as a cause of spring feeding by winter moth caterpillars. *Ecology, 51,* 565–581.

Fernandes, G. W., Almada, E. D. & Carneiro, M. A. A. (2010). Gall-inducing insect species richness as indicators of forest age and health. *Environmental Entomology, 39,* 1134–1140.

Fernandes, G. W. & Price, P. W. (1988). Biogeographical gradients in galling species richness: tests of hypotheses. *Oecologia, 76,* 161–167.

Fisher, R. A. (1930). *The general theory of natural selection.* First edition. Oxfordshire, Oxford: Oxford University Press.

Floate, K. D. & Whitham, T. G. (1993). The "hybrid bridge" hypothesis: Host shifting via plant hybrid swarms. *The American Naturalist, 4,* 651–652.

Floren, A. & Linsenmair, K. E. (2001). The influence of anthropogenic disturbances on the structure of arboreal arthropod communities. *Plant Ecology, 153,* 153-167.

Folliot, R. (1977). Les insectes cecidogenes et la cecidogenese. En: P. P. Grasse, editor. *Traité de Zoologie* (vol.8, pp. 389–429). Paris, France: Fasc. V B. Masson. [Folliot, R. (1977). Cecidogenic insects and cecidogenesis. In: P. P. Grasse, editor. *Treaty of Zoology* (vol.8, pp. 389–429). Paris, France: Fasc. B. Masson].

Forkner, R. E. & Hunter, M. D. (2000). What goes up must come down? Nutrient and predation pressure on oak herbivores. *Ecology, 81,* 1588–1600.

Forkner, R. E. Marquis, R. J. & Lill, J. T. (2004). Feeny revisited: condensed tannins as anti-herbivore defenses in leaf-chewing herbivore communities of *Quercus. Ecological Entomology, 29,* 174–187.

Forkner, R. E., Marquis, R. J., Lill, J. T. & Le Corff, J. (2006). Impacts of alternative timber harvest practices in leaf-chewing herbivores of oak. *Conservation Biology, 20,* 429–440.

Forkner, R. E., Marquis, R. J., Lill, J. T. & Corff, J. L. (2008). Timing is everything? Phenological synchrony and population variability in leaf-chewing herbivores of *Quercus. Ecological Entomology, 33,* 276–285.

Franks, A. J., Yates, C. J. & Hobbs, R. J. (2009). Defining plant functional groups to guide rare plant management. *Plant Ecology 204,* 207–216.

Freeman, J. S., O'Reilly-Wapstra, J. M., Vaillancourt, R. E., Wiggins, N. & Potts, B. M. (2008). Quantitative trait loci for key defensive compounds affecting herbivory of eucalypts in Australia. *New Phytologist, 178,* 846–851.

Frewen, B. E., Chen, T. H., Howe, G. T., Davis J., Rhode A., Boerjan W. & Bradash, H. D. Jr. (2000). Quantitative trait loci and candidate gene mapping of bud set and bud flush in *Populus. Genetics, 154,* 837–845.

Fritz, R. S. (1999). Resistance of hybrid plants to herbivores: genes, environment, or both? *Ecology, 80,* 382–391.

Fritz, R. S., Nichols-Orians, C. M. & Brunsfeld, S. J. (1994). Interespecific hybridization of plants and resistance to herbivores: hypotheses, genetics, and variable responses in a diverse community. *Oecologia, 97,* 106–117.

Fritz, R. S., Roche, B. M., Brunsfeld, S. J. & Orians, C. M. (1996). Interespecific and temporal variation in herbivores responses to hybrid willows. *Oecologia, 108,* 121–129.

Fukami, T. & Wardle, D. A. (2005). Long-term ecological dynamics: reciprocal insights from natural and anthropogenic gradients. *Proceedings of the Royal Society of London. Series B, 272,* 2105–15.

Fukui, A. (2001). Indirect interactions mediated by leaf shelters in animal–plant communities. *Population Ecology, 43,* 31–40.

Futuyma, D. J. & Agrawal, A. A. (2009). Macroevolution and the biological diversity of plants and herbivores. *Proceedings of the Royal Society of London. Series B, 106,* 18054–18061.

Gardner, T. (2010). *Monitoring Forest Biodiversity: Improving Conservation through Ecologically-Responsible Management.* First edition. London, England: Earthscan.

Gaston, K. J. (1991). Regional numbers of insect and plant species. *Functional Ecology, 6,* 243–247.

Gaston, K. J. (2000). Global patterns in biodiversity. *Nature, 405,* 220–227.

Gaston, K. J. & Lawton, J. H. (1988). Patterns in the distribution and abundance of insect populations. *Nature, 331,* 709–712.

Geber, M. A. & Griffen, L. R. (2003). Inheritance and natural selection on functional traits. *International Journal of Plant Sciences, 164,* S21–S42.

Genung, M. A., Schweitzer, J. A, Ubeda, F., Fitzpatrick, B. M., Pregitzer, C. C., Felker-Quinn, E. & Bailey, J. K. (2011). Genetic variation and community change: selection, evolution, and feedbacks. *Functional Ecology, 25,* 408–419.

Gering, J. C., Veech, J. A. & Crist, T. O. (2003). Additive partitioning of species diversity across multiple spatial scales: implications for regional conservation of biodiversity. *Conservation Biology, 17,* 488–499.

Gerisch, M., Agostinelli, V., Henle, K. & Dziock, F. (2012). More species, but all do the same: contrasting effects of flood disturbance on ground beetle functional and species diversity. *Oikos, 121,* 508–515.

Gibb, H., Johansson, T., Stenbacka, F. & Hjältén, J. (2013). Functional Roles Affect Diversity-Succession Relationships for Boreal Beetles. *PLoS ONE, 8,* e72764.

Gibson, G. A. P., Huber, J. T. & Woolley, J. B. (2006). *Annotated keys to the genera of Nearctic Chalcidoidea (Hymenoptera).* First edition. Ottawa, Canada: NRC Research Press.

Glynn, C., Rönnberg-Wästljun, A., Julkunen-Tiitto, R. and Weih, M. (2004). Willow genotype, but not drought treatment, affects foliar phenolic concentrations and leaf-beetle resistance. *Entomologia Experimentalis et Applicata,* 113, 1–14.

González-Rodríguez, A., Arias, D. M., Valencia, S. & Oyama, K. (2004). Morphological and RAPD analysis of hybridization between *Quercus affinis* and *Q. laurina* (Fagaceae), two Mexican red oaks. *American Journal of Botany,* 91, 401–409.

Gurevitch, J., Scheiner, S. M. & Fox, G. A. 2002. *The Ecology of Plants.* Second Edition. Sunderland: Sinauer.

Grimaldi, D. & Engel, M. S. (2005). *Evolution of the insects.* First edition. Cambridge, United Kingdom: Cambridge University Press.

Grytnes, J. A. & McCain C. M. (2007). Elevational trends in biodiversity. In: S. A. Levin, editor. *Encyclopedia of Biodiversity* (first edition, pp. 1–8). New York, USA: Elsevier Inc.

Haddad, N. M., Crutsinger, G. M., Gross, K., Haarstad, J., Knops, J. M. H. & Tilman, D. (2009). Plant species loss decreases arthropod diversity and shifts trophic structure. *Ecology Letters, 12,* 1029–39.

Haddad, N. M., Tilman, D., Haarstad, J., Ritchie, M. & Knops, J. M. H. (2001). Contrasting effects of plant richness and composition on insect communities: a field experiment. *The American Naturalist, 158,* 17–35.

Hairston, N. G., Smith, F. E. & Slobodkin, L. B. (1960). Community structure, population control, and competition. *The American Naturalist, 94,* 421–425.

Halaj, J., Ross, D. W. & Moldenke, A. R. (2000). Importance of habitat structure to the arthropod food-web in Douglas fir canopies. *Oikos, 90,* 139–152.

Hanski, I. (1994). Patch-occupancy dynamics in fragmented landscapes. *Trends in Ecology & Evolution, 9,* 131–135.

Hamer, K. C., Hill, J. K., Benedick, S., Mustaffa, N., Sherratt, T. N., Maryati, M. & Chey, V. K. (2003). Ecology of butterflies in natural and selectively logged forests of northern Borneo: the importance of habitat heterogeneity. *Journal of Applied Ecology, 40,* 150–162.

Hamilton, A. J., Basset, Y., Benke, K. K., Grimbacher, P. S., Miller, S. A., Novotny, V., Samuelson, G. A., Stork, N. E., Weiblen, G. D. & Yen, J. D. L. (2010). Quantifying uncertainty in estimation of tropical arthropod species richness. *The American Naturalist, 176,* 90–95.

Hardiman, B. S., Bohrer, G., Gough, C. M., Vogel, C. S. & Curtis, P. S. (2011). The role of canopy structural complexity in wood net primary production of a maturing northern deciduous forest. *Ecology, 92,* 1818–1827.

Harper, J. L. (1977). *Population Biology of Plants.* First edition. New York, New York, USA: Academic Press.

Hartley, S. E. (1998). The chemical composition of plant galls: are levels of nutrients and secondary compounds controlled by the gall-former? *Oecologia, 113,* 492–501.

Hawkins, B. A. & Porter, E. E. (2003). Does herbivore diversity depend on plant diversity? The case of California butterflies. *The American Naturalist, 161,* 40–49.

Hespenheide, H. A. (1991). Bionomics of leaf-mining insects. *Annual Review of Entomology, 36,* 535–560.

Hill, J. K. K., Hamer, K. C., Lace, L. A. & Banham, M. T. (1995). Effects of selective logging on tropical forest butterflies on Buru, Indonesia. *Journal of Applied Ecology, 32,* 754–760.

Hillebrand, H. (2004). On the generality of the latitudinal diversity gradient. *The American Naturalist, 163,* 192–211.

Hirao, T., Murakami, M., Kashizaki, A. & Ichtanabe, S. (2007). Additive apportioning of lepidopteran and coleopteran species diversity across spatial and temporal scales cool-temperate deciduous forest in Japan. *Ecological Entomology, 32,* 627–636.

Hochwender, C. G. and Fritz, R. S. (2004). Plant genetic differences influence herbivore community structure: evidence from a hybrid willow system. *Oecologia, 138,* 547–557.

Hodkinson, I. D. (2005). Terrestrial insects along elevation gradients: species and community responses to altitude. *Biological Reviews, 80,* 489–513.

Hooper, D. U., Chapin, F. S. III, Ewel, J. J., Hector, A., Inchausti, P., Lavorel, S., Lawton, J. H., Lodge, D. M., Loreau, M., Naeem, S., Schmid, B., Setälä, H., Symstad, A. J., Vandermeer, J. & Wardle, D. A. (2005). Effects of biodiversity on ecosystem functioning: a consensus of current knowledge. *Ecological Monographs, 75,* 3–35.

Hughes, A. R., Byrnes, J. E., Kimbro, D. L. & Stachowicz, J. J. (2007). Reciprocal relationships and potential feedbacks between biodiversity and disturbance. *Ecology Letters, 10,* 849–864.

Hughes, R., Inouye, B. D., Johnson, M. T. J., Underwood, N. & Vellend, M. (2008). Ecological consequences of genetic diversity. *Ecology Letters, 11,* 609–623.

Hunt, T., Bergsten, J., Levkanicova, Z., Papadopoulou, A., St. John, O., Wild, R., Hammond, P. M., Ahrens, D., Balke, M., Caterino, M. S., Gómez-Zurita, J., Ribera, I., Barraclough, T. G., Bocakova, M., Bocak, L. & Vogler, A. P. (2007). A comprehensive phylogeny of beetles reveals the evolutionary origins of a duperradiation. *Science, 318,* 1913–1916.

Hunter, M. D. & Price, P. W. (1992). Playing chutes and ladders: heterogeneity and the relative roles of bottom-up and top-down forces in natural communities. *Ecology, 73,* 723–732.

Hunter, M. D., Varley, G. C. & Gradwell, G. R. (1997). Estimating the relative roles of top-down and bottom-up forces on insect herbivore populations: A classic study revisited. *Proceedings of the National Academy of Sciences of the United States of America, 94,* 9176–9181.

Hunter, M. D. & Willmer, P.G. (1989). The potential for interspecific competition between two abundant defoliators on oak: leaf damage and habitat quality. *Ecological Entomology, 14,* 267–277.

Hutchinson, G. E. (1959). Homage to Santa Rosalia or why are there so many kinds of animals. *The American Naturalist, 93,* 145–159.

Ilg, C. Dziock, F., Foeckler, F., Follner, K., Gerisch, M., Glaeser, J., Rink, A., Schanowski, A., Scholz, M., Deichner, O. & Henle, K. (2008). Long-term reactions of plants and macroinvertebrates to extreme floods in floodplain grasslands. *Ecology, 89,* 2392–2398.

Inoue, T. (2003). Chronosequential change in a butterfly community after clear-cutting of deciduous forests in a cool temperate region of central Japan. *Entomological Science, 6,* 151–163.

Ishii, H. T., Tanabe, S. & Hiura, T. (2004). Exploring the relationships among canopy structure, stand productivity and biodiversity of temperate forest ecosystems. *Journal of Forest Science, 50,* 342–355.

Janz, N. (2011). Ehrlich and Raven revisited: mechanisms underlying codiversification of plants and enemies. *Annual Review of Ecology, Evolution, and Systematics, 42,* 71–89.

Jeffries, J. M., Marquis, R. J. and Forkner, R.E. (2006). Forest age influences oak insect herbivore community structure, richness and density. *Ecological Applications, 16,* 901–912.

Jermy, T. (1984). Evolution on insect/host plant relationship. *The American Naturalist, 124,* 609–630.

Johnson, M. T. J. & Agrawal, A. A. (2005). Plant genotype and the environment interact to shape a diverse arthropod community on Evening Primrose (*Oenothera biennis*). *Ecology, 86,* 874–875.

Johnson, M. T. J., Lajeunesse, M. J. & Agrawal, A. A. (2006). Additive and interactive effects of plant genotypic diversity on arthropod communities and plant fitness. *Ecology Letters, 9,* 24–34.

Kallimanis, A. S., Mazaris, A. D., Tzanopoulos, J., Halley, J. M., Pantis, J. D. & Sgardelis, S. P. (2008). How does habitat diversity affect the species-area relationship? *Global Ecology and Biogeography, 17,* 532–538.

Kause, A., Ossipov, V., Haukioja, E., Lempa, K., Hanhimaki, S. & Ossipova, S. (1999). Multiplicity of biochemical factors determining quality of growing birch leaves. *Oecologia, 120,* 102–112.

Keim, P., Paige, K. N., Whitham, T. G. & Lark, K. G. (1989). Genetic analysis of an interespecific hybrid swarm of *Populus*: occurrence of unidireccional introgression. *Genetics, 123,* 557–565.

Kessler, M., Kluge, J., Hemp, A. & Ohlemuller, R. (2011). A global comparative analysis of elevational species richness patterns of ferns. *Global Ecology and Biogeography, 20,* 868–880.

Kinsey, A. C. (1930). The gall wasp genus Cynips. A study in the origin of species. *Indiana University Studies, 84–86,* 1–577.

Knoll, A. H., Niklas, K. J. & Tiffney, B. H. (1979). Phanerozoic land-plant diversity in North America. *Science, 206,* 1400–1402.

Knops, J. M. H., Tilman, D., Haddad, N. M., Naeem, S., Mitchell, C. E., Haarsted, J., Ritchie, M. E., Howe, K. M., Reich, P. B., Siemann, E. & Groth, J. (1999). Effects of plant species richness on invasion dynamics, disease outbreaks, insect abundances and diversity. *Ecology Letters, 2,* 286–293.

Körner, C. (2007). The use of 'altitude' in ecological research. *Trends in Ecology & Evolution, 22,* 569–74.

Kraft, N. J. B., Comita, L. S., Chase, J. M., Sanders, N. J., Swenson, N. G., Crist, T. O. Stegen, J. C., Vellend, M., Boyle, B., Anderson, M. J., Cornell, H. V., Davies, K. F., Freestone, A. L., Inouye, B. D., Harrison, S. P. & Myers, J. A. (2011). Disentangling the drivers of diversity along latitudinal and elevational gradients. *Science, 333,* 1755–1758.

Kumar, S., Simonson, S. & Stohlgren, T. J. (2009). Effects of spatial heterogeneity on butterfly species richness in Rocky Mountain National Park, CO, USA. *Biodiversity and Conservation, 18,* 739–763.

Kursar, T. A. & Coley, P. D. (2003). Convergence in defense syndromes of young leaves in tropical rainforest. *Biochemical Systematics and Ecology, 21,* 929–949.

Labandeira C. C. (2006). The four phases of plant-arthropod associations in deep time. *Geologica Acta,* 4: 409–438.

Labandeira, C. C., Dilcher, D. L., Davis, D. R. & Wagner, D. L. (1994). Ninety-seven million years of angiosperm-insect association: Paleobiological insights into the meaning of coevolution. *Proceedings of the National Academy of Sciences of the United States of America, 91,* 12278–12282.

Labandeira, C. C. & Sepkoski, J. J. (1993). Insect diversity in the fossil record. *Science, 261,* 310–315.

Lambert, L., McPherson, R. M. & Espelie, K. E. (1995). Soybean host plant resistance mechanisms that alter abundance of white-flies (Homoptera: Alyrodidae). *Environmental and Ecological Statistics, 24,* 1381–1386.

Larsson, S., Haggstrom, H. & Denno, R. F. (1997). Preference for protected feeding sites by larvae of the willow-feeding leaf beetle *Galerucella lineola. Ecological Entomology, 22,* 445–452.

Larsson, S. I., Wiren, A. I., Lundgren, L. I. & Ericsson, I. (1986). Effects of light and nutrients stress on leaf phenolic chemistry in *Salix dasyclados* and susceptibility *Galerucella lineola* (Coleoptera). *Oikos, 47,* 205–210.

Lawton, J. H. & Strong, D. R. Jr. (1981). Community patterns and competition in folivorous insects. *The American Naturalist, 118,* 317–38.

Le Corff, J. & Marquis, R. J. (1999). Difference between understory and canopy in herbivore community composition and leaf quality for two oak species in Missouri. *Ecological Entomology, 24,* 46–58.

Lessard, J. P., Sackett, T. E., Reynolds, W. N., Fowler, D. A. & Sanders, N. J. (2011). Determinants of the detrital arthropod community structure: the effects of temperature and resources along an environmental gradient. *Oikos, 120,* 333–343.

Lewis, O. T. (2001). Effects of experimental selective logging on tropical butterflies. *Conservation Biology, 15,* 389–400.

Lill, J. T. & Marquis, R. J. (2001). The effects of leaf quality on herbivore performance and attack from natural enemies. *Oecologia, 126,* 418–428.

Lill, J. T. & Marquis, R. J. (2003). Ecosystem engineering by caterpillars increases insect herbivore diversity on white oak. *Ecology, 84,* 682–690.

Lindeman, R. L. (1942). The trophic-dynamic aspect of ecology. *Ecology, 23,* 399–408.

Lomolino, M. V., Sax, D. F., Riddle, B. R. & Brown, J. H. (2006b). The island rule and a research agenda for studying ecogeographical patterns. *Journal of Biogeography, 33,* 1503–1510.

McCain, C. M. & Colwell, R. K. (2011). Assessing the threat to montane biodiversity from discordant shifts in temperature and precipitation in a changing climate. *Ecology Letters, 14,* 1236–1245.

Mackey, R. L. and Currie, D. J. (2001). The diversity–disturbance relationship: is it generally strong and peaked? *Ecology, 82,* 3479–3492.

Maldonado-López, Y., Cuevas-Reyes, P., Stone, G. N., Nieves-Aldrey, J. L. & Oyama, K. (2015). Gall wasp community response to fragmentation of oak tree species: importance of fragment size and isolated trees. *Ecosphere, 6,* 1–15.

Maldonado-López, Y., Espinoza-Olvera, N. A., Pérez-López, G., Quesada-Béjar, V., Oyama, K., González-Rodríguez, A. & Cuevas-Reyes, P. (2013). Interacciones antagónicas especialistas en encinos: El caso de los insectos inductores de agallas. *Biológicas, 2,* 31–41. [Maldonado-Lopez, Y., Espinoza-Olvera, N. A., Perez-Lopez, G., Quesada-Béjar, V., Oyama, K., Gonzalez-Rodriguez, A. & Cuevas-Reyes, P. (2013). Specialist antagonistic interactions in oaks: The case of gall-inducing insects. *Biologicas, 2,* 31–41].

Marquis, R. J. & Le Corff, J. (1997). Estimating pretreatment variation in the oak leaf chewing insect fauna of the Missouri Ozark Forest Ecosystem Project (MOFEP). In: B. Brookshire, B. & S. Shifley (Eds.), *Proceedings of the Missouri Ozark Forest Ecosystem Project Symposium, GTRNC-193* (first edition, pp. 332–346) North Central

Experiment Station, St. Paul, Minnesota, USA: Department of Agriculture, Forest Service.

Marquis, R. J. & Lill, J. T. (2010). Impact of plant architecture versus leaf quality on attack by leaf-tying caterpillars on five oak species. *Oecologia, 163*, 203–213.

Marquis, R. J., Lill, J. T. & Piccini, A. (2002). Effect of plant architecture on colonization and damage by leaftying caterpillars of *Quercus alba. Oikos, 99*, 531–537.

Martinsen, G. D., Floate, K. D., Waltz, A. M., Wimp, G. M. & Whitham, T. G. (2000). Positive interactions between leafrollers and other arthropods enhance biodiversity on hybrid cottonwoods. *Oecologia, 123*, 82–89.

Martinsen, G. D. & Whitham, T. G. (1994). More birds nest in hybrid cottonwoods. *Wilson Bulletin, 106*, 474–481.

Mattson Jr, W. J. (1980). Herbivory in relation to plant nitrogen content. *Annual Review of Ecology, Evolution, and Systematics, 11*, 119–161.

Mayhew, P. J. (2007). Why are there so many insect species? Perspectives from fossils and phylogenies. *Biological Reviews, 82*, 425–454.

McArt, S. H., Cook-Patton S. C. & Thaler, J. S. (2012). Relationships between arthropod richness, evenness, and diversity are altered by complementarity among plant genotypes. *Oecologia, 168*, 1013–1021.

McCain, C. M. & Colwell, R. K. (2011). Assessing the threat to montane biodiversity from discordant shifts in temperature and precipitation in a changing climate. *Ecology Letters, 14*, 1236–1245.

McKinney, M. L. (2008). Effects of urbanization on species richness: A review of plants and animals. *Urban Ecosystems, 11*, 161–176.

Menke, S. B. & Holway, D. A. (2006). Abiotic factors control invasion by ants at the community scale. *Journal of Animal Ecology, 75*, 368–76.

Mitter, C., Farrell, B. D. & Wiegmann, B. (1988). The phylogenetic study of adaptive zones. Has phytophagy promoted insect diversification? *The American Naturalist, 132*, 107–128.

Morais, J. M. & Cianciaruso, M. (2014). Plant functional groups: scientometric analysis focused on removal experiments. *Acta Botanica Brasílica, 28*, 502–511.

52   *E. Tovar-Sánchez, E. Castillo-Mendoza, L. Valencia-Cuevas et al.*

Moreau, C. S., Bell, C. D., Vila, R., Archibald, S. B. & Pierce, N. E. (2006). Phylogeny of the ants: diversification in the age of angiosperms. *Science, 312,* 101–104.

Murakami, M., Hirao, T. & Ichie, T. (2007). Comparison of lepidopteran larval communities among tree species in a temperate deciduous forest, Japan. *Ecological Entomology, 32,* 613–620.

Murakami, M., Ichie, T. & Hirao, T. (2008). Beta-diversity of lepidopteran larval communities in a Japanese temperate forest: effects of phenology and tree species. *Ecological Research, 23,* 173–189.

Murakami, M., Yoshida, K., Hara, H. & Toda, M. J. (2005). Spatiotemporal variation in lepidopteran larval assemblages associated with oak *Quercus crispula*: the importance of leaf quality. *Ecological Entomology, 30,* 521–531.

Nakamura, T., Hattori, K., Ishida, T. A., Sato, H. & Kimura, M. T. (2008). Population dynamics of leafminers on a deciduous oak *Quercus dentata. Acta Oecologica, 34,* 259–265.

Nason, J. D. (2002). La estructura genética de las poblaciones de árboles. En: M. R. Guariguata & G. H. Kattan (Eds.), *Ecología y Conservación de los Bosques Neotropicales* (primera edición, pp. 299–327). Costa Rica, Ediciones LUR. [Nason, J. D. (2002). The genetic structure of tree populations. In: M. R. Guariguata & G. H. Kattan (Eds.), *Ecology and Conservation of Neotropical Forests* (first edition, pp. 299–327). Costa Rica, Ediciones LUR].

Nazemi, J., Talebi, A. A., Sadeghi, S. E., Melika, G. & Lozan, A. (2008). Species richness oak wasps (Hymenoptera: Cynipidae) and identification of associated inquilines and parasitoids on two oak species in western Iran. *Norwegian Journal of Zoology, 4,* 189–202.

Newcombe, G. & Bradshaw, H. D. Jr. (1996). Quantitative trait loci conferring resistance in hybrid poplar to *Septoria populicola*, the cause of leaf spot. *Canadian Journal Forest Research, 26,* 1943–1950.

Niemelä, J., Kotze, D. J., Venn, S., Penev, L., Stoyanov, I., Spence, J., Hartley & D., Montes de Oca, E. (2002). Carabid beetle assemblages (Coleoptera, Carcabidae) across urban-rural gradients: an international comparison. *Landscape Ecology, 17,* 387–401.

Nierenstein, M. (1930). Interrelation between gallproducers and galls. *Nature, 125*, 348–342.

Nieves-Aldrey, J. L. (1987). Estado actual del conocimiento de la subfamilia Cynipinae (Hymenoptera, Parasitica, Cynipidae) en la Peninsula Iberica. *Eos, 63*, 179–195. [Nieves-Aldrey, J. L. (1987). Current knowledge status of the Cynipinae subfamily (Hymenoptera, Parasitica, Cynipidae) in the Iberian Peninsula. *Eos, 63*, 179–195].

Novotny, V., Drozd, P., Miller, S. E., Kulfan, M., Janda, M., Basset, Y. & Weiblen, G. D. (2006). Why are there so many species of herbivorous insects in tropical rainforests? *Science, 313*, 1115–1118.

Olson, D. M. (1994). The distribution of leaf litter invertebrates along a neotropical altitudinal gradient. *Journal of Tropical Ecology, 10*, 129–150.

Oyama, K. (1999). La Coevolución. En: J. Núñez-Farfán & E. Eguiarte (Eds.), *La Evolución Biológica* (primera edición, pp. 153–171). México City: Universidad Nacional Autónoma de México. [Oyama, K. (1999). The Coevolution. In: J. Núñez-Farfán & E. Eguiarte (Eds.), *The Biological Evolution* (first edition, pp. 153–171). México City: National Autonomous University of Mexico].

Oyama, K. (2012). Coevolución. In: E. Del Val & K. Boege (Eds.), *Ecología y Evolución de las interacciones bióticas* (primera edición, pp. 204–226). FCE, UNAM, IE. Ciudad de México. [Coevolution. In: E. Del Val & K. Boege (Eds.), *Ecology and evolution of biotic interactions* (first edition, pp. 204–226). FCE, UNAM, IE. Mexico, City.

Palacios-Vargas, J. G., Iglesias, R. & Castaño-Meneses, G. (2003). Mites from Mexican canopies. *Insect Science and Its Application, 23*, 287–292.

Pellissier, L., Fiedler, K., Ndribe, C., Dubuis, A., Pradervand, J. N., Guisan, A. & Rasmann, S. (2012). Shifts in species richness, herbivore specialisation and plant resistance along elevation gradients. *Ecology and Evolution, 2*, 1818–1825.

Pénzes, Z., Tang, C. T., Bihari, P., Bozsó, M., Schwéger, S. & Melika, G. (2012). Oak associated inquilines (Hymenoptera, Cynipidae, Synergini). *Tiscia Monograph, 11*, 76.

Porter, W. P. & Gates, D. M. (1969). Thermodynamic equilibria of animals with environment. *Ecological Monographs, 39,* 227–244.

Preszler, R. W. & Boecklen, W. J. (1994). A tree-trophic-level analysis of the effects of plants hybridization on a leaf-mining moth. *Oecologia, 100,* 66–73.

Pujade-Villar, J. (2013). Las agallas de los encinos: un ecosistema en miniatura que hace posibles estudios multidiciplinares. *Folia EntomológicaMexicana, 12,* 1–20. [Pujade-Villar, J. (2013). The galls of the oaks: a miniature ecosystem that makes possible multidisciplinary studies. *Mexican Entomological Folia, 12,* 1–20].

Rahbek, C. (2005). The role of spatial scale and the perception of largescale species-richness pattern. *Ecology Letters, 8,* 224–339.

Rainford, J. L., Hofreiter, M., Nicholson, D. B. & Mayhew, P. J. (2014). Phylogenetic distribution of extant richness suggests metamorphosis is a key innovation driving diversification in insects. *PLoS One, 9,* e109085.

Ricklefs, R. E. (2008). Foliage chemistry and the distribution of Lepidoptera larvae on broad-leaved trees in southern Ontario. *Oecologia, 157,* 53–67.

Rieseberg, L. H. & Ellstrand, N. C. (1993). What can molecular and morphological markers tell us about plant hybridization? *Critical Reviews in Plant Sciences, 12,* 213–241.

Rivera, G. E. (1991). Métodos y técnicas para determinar el régimen alimenticio en insectos herbívoros. *Boletín de la Sociedad Mexicana de Entomología, 8,* 27–33. [Rivera, G. E. (1991). Methods and techniques to determine the diet in herbivorous insects. *Bulletin of the Mexican Society of Entomology, 8,* 27–33].

Rohdendorf B. B. & Raznitsin A. P. (1989). *The historical development of the class Insecta.* Moscu: Trudy Paleontology Institute.

Rodríguez-Castañeda, G., Dyer, L. A., Brehm, G., Connahs, H., Forkner, R. E. & Walla, T. R. (2010). Tropical forests are not flat: how mountains affect herbivore diversity. *Ecology Letters, 13,* 1348–1357.

Rodríguez-Domínguez, A. (2010). *Efecto de un gradiente altitudinal sobre la estructura de la comunidad de artrópodos asociados a hojarasca en el bosque de Abies-Quercus en el parque nacional El Chico, Hidalgo, México*. Tesis de Licenciatura. Cuernavaca, Morelos, México: Universidad Autónoma del estado de Morelos. [Rodríguez-Domínguez, A. (2010). *Effect of an altitudinal gradient on arthropod community structure associated of leaf litter in the Abies-Quercus forest in El Chico National Park, Hidalgo, Mexico*. Bachelor thesis. Cuernavaca, Morelos, México: Autonomous University of Morelos State.

Ronquist, F. (1994). Evolution of parasitism among closely related species: phylogenetic relationships and the origin of inquilinism in gall wasps (Hymenoptera, Cynipidae). *Evolution, 48,* 241–266.

Ronquist, F. (1999). Phylogeny, classification and evolution of the Cynipoidea. *Zoologica Scripta, 28,* 139–164.

Ronquist, F. & Liljeblad, J. (2001). Evolution of the gall wasp–host plant association. *Evolution, 55,* 2503–2522.

Rzedowski, J. (1978). *Vegetación de México*. Primera edición. Ciudad de México, México: Limusa. [Rzedowski, J. (1978). *Vegetation of Mexico*. First edition. Mexico City, Mexico: Limusa].

Salminen, J. P., Roslin, T., Karonen, M., Sinkkonen, J., Pihlaja, K. & Pulkkinen, P. (2004). Seasonal variation in the content of hydrolyzable tannins, flavonoid glycosides, and proanthocyanidins in oak leaves. *Journal of Chemical Ecology, 30,* 1693–1711.

Sanders, N. J., Moss, J. & Wagner, D. (2003). Patterns of ant species richness along elevational gradients in an arid ecosystem. *Global Ecology and Biogeography, 12,* 93–102.

Sarfraz, M., Dosdall, L. M. & Keddie, B. A. (2008). Host plant genotype of the herbivore *Plutella xylostela* (Lepidopetera: Plutellidae) affects the performance of its parasitoid *Diadegma insulare* (Hymenoptera: Ichneumonidae). *Biological Control, 44,* 42–51.

Schädler, M., Jung, G., Auge, H. & Brandl, R. (2003). Does the Fretwell–Oksanen model apply to invertebrates? *Oikos, 100,* 203–207.

Sheley, R. L. & J. J. James (2010). Resistance of Native Plant Functional Groups to Invasion by Medusahead (*Taeniatherum caput-medusae*). *Invasive Plant Science and Management, 3,* 294–300.

Scherber, C., Eisenhauer, N., Weisser, W. W., Schmid, B., Voigt, W., Fischer, M., Schulze, E. D., Roscher, C., Weigelt, A., Allan, E., Beszler, H., Bonkowski, M., Buchmann, N., Buscot, F., Clement, L. W., Ebeling, A., Engels, C., Halle, S., Kertscher, I., Klein, A. M., Koller, R., Konig, S., Kowalski, E., Kummer, V., Kuu, A., Lange, M., Lauterbach, D., Middelhoff, C., Migunova, V.D., Milcu, A., Muller, R., Partsch, S., Petermann, J. S., Renker, C., Rottstock, T., Sabais, A., Scheu, S., Schumacher, J., Temperton, V. M. & Tscharntke, T. (2010). Bottom-up effects of plant diversity on multitrophic interactions in a biodiversity experiment. *Nature, 468,* 553–556.

Schluter, D. (2001). Ecology and the origin of species. *Trends in Ecology & Evolution, 16,* 372–380.

Schönrogge, K., Harper, L. J. & Lichtenstein, C. P. (2000). The protein content of tissues in cynipid galls (Hymenoptera: Cynipidae): similarities between cynipid galls and seeds. *Plant, Cell & Environment, 23,* 215–22.

Schoonhoven, L. M., Van Lonn, J. J. A. & Dicke, M. (2005). *Insect-plant biology.* Second edition. Oxfordshire, Oxford: Oxford University Press.

Schowalter, T. D. (1995). Canopy arthropod community response to forest age and alternative harvest practices in western Oregon. *Forest Ecology and Management, 78,* 115–25.

Schowalter, T. D. (2000). *Insect ecology. An ecosystem approach.* First edition. London, United Kingdom: Academic Press.

Schowalter, T. D. (2011). *Insect ecology: an ecosystem approach.* Third edition. San Diego, California, USA: Academic Press.

Shepherd, M., Chaparro, J. X. & Teasdale, R. (1999). Genetic mapping of monoterpene composition in an interspecific eucalypt hybrid. *Theoretical and Applied Genetics, 99,* 1207–1215.

Shuster, S. M., Lonsdorf, E. V., Wimp, G. M., Bailey, J. K. & Whitham, T. G. (2006). Community heritability measures the evolutionary consequences of indirect genetic effects on community structure. *Evolution, 60,* 991–1003.

Şen, I. & Gök, A. (2009). Leaf beetle communities (Coleoptera: Chrysomelidae) of two mixed forest ecosystems dominated by pine–

oak–hawthorn in Isparta province, Turkey. *Annales Zoologici Fennici, 46*, 217–232.

Serrano-Muñoz, M. (2016). *Diversidad de cinípinos (Hymenoptera: Cynipidae) y de himenópteros (Synergini y Chalcidoidea) asociados a agallas de encinos de la región noroeste de la Sierra de Guadalupe.* Tesis de maestría. Ciudad de México, México: Instituto Politécnico Nacional. [Serrano-Muñoz, M. (2016). *Diversity of cynipid (Hymenoptera: Cynipidae) and Hymenoptera (Synergini and Chalcidoidea) associated to oak galls of the northwest region of the Sierra de Guadalupe.* Master's Thesis. Mexico City, Mexico: National Polytechnic Institute].

Sheley, R. L. & James, J. (2010). Resistance of Native Plant Functional Groups to Invasion by Medusahead (Taeniatherum caput-medusae). *Invasive Plant Science and Management*, 3: 294-300.

Siemann, E. (1998). Experimental tests of effects of plant productivity and diversity on grassland arthropod diversity. *Ecology, 79*, 2057-2070.

Siemann, E., Tilman, D., Haarstad, J. & Ritchie, M. (1998). Experimental test of the dependence of arthropod diversity on plant diversity. *The American Naturalist, 152*, 738–750.

Singer, M. S. & Stireman, J. O. (2005). The tri-trophic niche concept and adaptive radiation of phytophagous insects. *Ecology Letters, 8,* 1247–1255.

Sobek, S., Steffan-Dewenter, I., Scherber, C. & Tschamtke, T. (2009). Spatio temporal changes of beetle communities across a tree diversity gradient. *Diversity & Distributions, 15*, 660–670.

Soumela, J. & Ayres, M. J. (1994). Wtihin-tree and among-tree variation in leaf characteristics of mountain birch and its implications for herbivory. *Oikos, 70,* 121–222.

Southwood, T. R. E., Wint, G. R. W., Kennedy, C. E. J. & Greenwood, S. R. (2004). Seasonality abundance, species richness and specificity of the phytophagous guild of insects on oak (*Quercus*) canopies. *European Journal of Entomology, 101,* 43–50.

Southwood, T. R. E., Wint, G. R. W., Kennedy, C. E. J. & Greenwood, S. R. (2005). The composition of the arthropod fauna of the canopies of some species of oak (*Quercus*). *European Journal of Entomology, 102,* 65–72.

Sporne, K. R. (1975). *The Morphology of Pteridophites.* Fourth edition. Hutchinson, London: Hillary House.

Srivastava, D. & Lawton, J. (1998). Why more productive sites have more species: an experimental test of theory using tree-hole communities. *The American Naturalist, 152,* 510–529.

Stein, A., Gerstner, K. and Kreft, H. (2014). Environmental heterogeneity as a universal driver of species richness across taxa, biomes and spatial scales. *Ecology Letters, 17,* 866–880.

Stone, G. N. & Cook, J. M. (1998). The structure of cynipid oak galls: patterns in the evolution of an extended phenotype. *Proceedings Biological Sciences, 265,* 979–988.

Stone, G. N. & Schönrogge, K. (2003). The adaptive significance of insect gall morphology. *Trends in Ecology & Evolution, 18,* 512–522.

Stone, G. N., Schönrogge, K., Atkinson, R. J., Bellido, D. & Pujade-Villar, J. (2002). The population biology of oak gall wasps (Hymenoptera: Cynipidae). *Annual Review of Entomology, 47,* 633–668.

Strauss, S. Y. (1994). Levels of herbivory and parasitism in host hybrid zones. *Trends in Ecology & Evolution, 9,* 209–214.

Strong, D. R. Jr, Lawton, J. H. & Southwood, T. R. E. (1984). *Insects on Plants: Community Patterns and Mechanisms.* First edition. Cambridge, MA: Harvard University Press.

Summerville, K. S. & Crist, T. O. (2002). Effects of timber harvest on forest Lepidoptera: community, guild, and species responses. *Ecological Applications, 12,* 820–835.

Summerville, K. S. & Crist, T. O. (2003). Determinants of lepidopteran species diversity and composition in eastern deciduous forest: roles of season, region and patch size. *Oikos, 100,* 134–148.

Sundqvist, M. K., Sanders, N. J. & Wardle, D. A. (2013). Mechanisms, and Insights for Global Change. *Annual Review of Ecology, Evolution, and Systematics, 44,* 261–80.

Tews, J., Brose, U., Grimm, V., Tielborger, K., Wichmann, M. C., Schwager, M. & Jeltsch, F. (2004). Animal species diversity driven by habitat heterogeneity/diversity: the importance of keystone structures. *Journal of Biogeography, 31,* 79–92.

Thomas, C. D., Singer, M. C., Mallet, J. L. B., Parmesan, C. & Billington, H. L. (1987). Incorporation of a European weed into the diet of a North American herbivore. *Evolution, 41,* 892–901.

Thompson, J. N. (2001). The Geographic Dynamics of Coevolution. In: C. W. Fox, D. A. Roff & D. F. Fairbairn (Eds.) *Evolutionary ecology. Concepts and case studies* (first edition, pp. 331–343). Albany, New York: Oxford University Press.

Thompson. J. N. (2013). *Relentless evolution.* First edition. Chicago, USA: University of Chicago Press.

Tilman, D., Wedin, D. & Knops, J. M. H. (1996). Productivity and sustainability influenced by biodiversity. *Nature, 379,* 718–20.

Tomas, F., Abbott, J. M., Balk, M., Steinberg, C., Williams, S. L. & Stachowicz, J. J. (2011). Plant genotype and nitrogen loading influence seagrass productivity, biochemistry, and plant-herbivore interactions. *Ecology, 92,* 1807–1817.

Tovar-Sánchez, E. (2009). Canopy arthropods community within and among oak species in central Mexico. *Acta Zoologica Sinica, 55,* 132–144.

Tovar-Sánchez, E., Cano-Santana, Z. & Oyama, K. (2003). Canopy arthropod communities on Mexican oaks at sites with different disturbance regimes. *Biological Conservation, 115,* 79–87.

Tovar-Sánchez, E. & Oyama, K. (2004). Natural hybridization and hybrid zones between *Quercus crassifolia* and *Quercus crassipes* (Fagaceae) in Mexico: morphological and molecular evidence. *American Journal of Botany, 91,* 1352–1363.

Tovar-Sánchez, E. & Oyama, K. (2006a). Community structure of canopy arthropods associated in *Quercus crassifolia* × *Quercus crassipes* complex. *Oikos, 112,* 370–381.

Tovar-Sánchez, E. & Oyama, K. (2006b). Effect of hybridization of the *Quercus crassifolia* x *Quercus crassipes* complex on the community structure of endophagous insects. *Oecologia, 147,* 702–713.

Tovar-Sánchez, E., Martí-Flores, E., Valencia-Cuevas, L. & Mussali-Galante, P. (2015a) Influence of forest type and host plant genetic relatedness on the canopy arthropod community structure of *Quercus crassifolia. Revista Chilena de Historia Natural, 88,* 7.

Tovar-Sánchez, E., Valencia-Cuevas, L., Castillo-Mendoza, E., Mussali-Galante, P., Pérez-Ruíz, R. V. & Mendoza, A. (2013). Association between individual genetic diversity of two oak host species and canopy arthropod community structure. *European Journal Forest Research, 132,* 165–179.

Tovar-Sánchez, E., Valencia-Cuevas, L., Mussali-Galante, P., Ramírez-Rodríguez, R. & Castillo-Mendoza, E. (2015b). Effect of host–plant genetic diversity on oak canopy arthropod community structure in central Mexico. *Revista Chilena de Historia* Natural, 88, 12.

Unsicker, S. B., Oswald, A., Kohler, G. & Weisser, W. W. (2008). Complementarity effects through dietary mixing enhance the performance of a generalist insect herbivore. *Oecologia, 156,* 313–324.

Valencia-Cuevas, L. & Tovar-Sánchez, E. (2015). Oak canopy arthropod communities: wich factors shape its structure? *Revista Chilena de Historia Natural, 88,* 1−22.

Valencia-Cuevas, L., Mussali-Galante, P., Cano-Santana, Z., Pujade-Villar, J., Equihua-Martínez, A. & Tovar-Sánchez, E. (2017). Genetic variation in foundation species governs the dynamics of trophic interactions. *Current Zoology, 64,* 13–22.

Vehviläinen, H., Koricheva, J. & Ruohomaki, K. (2008). Effects of stand tree species composition and diversity on abundance of predatory arthropods. *Oikos, 117,* 935–943.

Waltz, A. M. & Whitham, T. G. (1997). Plant development affects arthropod communities: opposing impacts of species removal. *Ecology, 78,* 2133–44.

White, P. S. & Pickett, S. T. A. (1985). Natural disturbance and patch dynamics: An Introduction. In: S. T. A. Pickett & P. S. White (Eds.), *The ecology of natural disturbance and patch dynamics.* (first edition, pp. 3–13). Orlando, Florida: Academic Press.

White, J. A. & Whitham, T. G. (2000). Associational susceptibility of cottonwood to a box elder herbivore. *Ecology, 81,* 1795–1803.

Whitham, T. G. (1989). Plant hybrid zones as sink for pests. *Science, 244,* 1490–1493.

Whitham, T. G., Bailey, J. K., Scheweitzer, J. A., Shuster, S. M., Bangert, R. K., LeRoy, C. J., Lonsdorf, E. V., Allan, G. J., DiFazio, S. P., Potts, B. M., Fischer, D. C., Gehring, C. A., Lindroth, R. L., Marks, J. C., Hart, S. C., Wimp, G. M. & Wooley, S. C. (2006). A framework for community and ecosystem genetics: form genes to ecosystems. *Nature, 7,* 510–523.

Whitham, T. G., Gehring, C. A., Lamit, L. J., Wojtowicz, T., Evans, L. M., Keith, A. R. & Smith, D. S. (2012). Community specificity: life and afterlife effects of genes. *Trends in Plant Science, 17,* 271–281.

Whitham, T. G., Martinsen, G. D., Floate, K. D., Dungey, H. S., Potts, B. M. & Keim, P. (1999). Plant hybrid zones affect biodiversity: tools for a genetic based understanding of community structure. *Ecology, 80,* 416–428.

Whitham, T. G., Morrow, P. A. & Potts, B. M. (1994). Plant hybrid zones as centers of biodiversity: the herbivore community of two endemic Tasmanian eucalypts. *Oecologia, 97,* 481–490.

Whitham, T. G., Young, W. P., Martinsen, G. D., Gebring, C. A., Scheweitzer, J. A., Shuster, S. M., Wimp, G. M., Fischer, D. C., Bailey, J. K., Lindroth, R. L., Woolbright, S. & Kuske, R. (2003). Community and ecosystem genetics: a consequence of the extended phenotype. *Ecology*, 84, 559–573.

Whitney, K. D., Ahern, J. R., Campbell, L. G., Albert, L. P. & King, M. S. (2010). Patterns of hybridization in plants. *Perspectives in Plant Ecology, Evolution and Systematics, 12,* 175–182.

Wiebes-Rijks, A. A. & Shorthouse, J. D. (1992). Ecological relationships of insects inhabiting cynipid galls. In: J. D. Shorthouse & O. Rohfritsch (Eds.), *Biology of Insect-induced Galls* (first edition, pp. 238–257). Albany, New York: Oxford University Press.

Willmer, P. G. (1982). Microclimate and the environmental physiology of insects. *Advances in Insect Physiology, 16,* 1–57.

Wilson, E. O. (1992). *The diversity of life*. First edition. Cambridge, MA: Harvard University Press.

Wimp, G. M., Martinsen, G. D., Floate, K. D., Bangert, R. K. & Whitham, T. G. (2005). Plant genetic determinants of arthropod community structure and diversity. *Evolution, 59,* 61–69.

Wimp, G. M., Young, P. W., Woolbright, S. A., Martinsen, G. D., Keim, P. & Whitham, T. G. (2004). Conserving plant genetic diversity for dependent animal communities. *Ecology Letters, 7,* 776–780.

Wimp, G. M., Wooley, S., Bangert, K., Young, W. P., Martinsen, G. D., Keim, P., Rehill, B., Lindroth, R. L. & Whitham, T. G. (2007). Plant genetics intra-annual variation in phytochemistry and arthropod community structure. *Molecular Ecology, 16,* 5057–5069.

Wojtowicz, T., Compson, Z. G., Lamit, L. J., Whitham, T. G. & Gehring. C. A. (2014). Plant genetic identity of foundation tree species and their hybrids affects a litter- dwelling generalist predator. *Oecologia, 176,* 799–810.

Wold, E. N. & Marquis, R. J. (1997) Induced defense in white oak: effects on herbivores and consequences for the plant. *Ecology, 78,* 1356–1369.

Wooton, R.J. (1981). Paleozoic Insects. *Annual Review of Entomology, 26,* 319-344.

Yarnes, C. T. & Boecklen, W. J. (2005). Abiotic factors promote plant heterogeneity and influence herbivore performance and mortality in Gambel's oak (*Quercus gambelii*). *Entomologia Experimentalis et Applicata, 114,* 87–95.

Yarnes, C. T., Boecklen, W. J. & Salminen, J. P. (2008). No simple sum: seasonal variation in tannin phenotypes and leafminers in hybrid oaks. *Chemoecology, 18,* 39–5.

In: Focus on Arthropods Research   ISBN: 978-1-53614-343-0
Editor: Mirko Messana   © 2018 Nova Science Publishers, Inc.

Chapter 2

INFERRING STAGES AND DEVELOPMENT TIMES OF EMBRYONIZED DECAPOD (PLEOCYEMATA) CRUSTACEAN LARVAE BASED ON EQUIVALENT FREE-LIVING STAGES OF DENDROBRANCHIATE PRAWNS

Brady K. Quinn[*]
Department of Biological Sciences, University of New Brunswick,
Saint John, NB, Canada

ABSTRACT

Many species of decapod crustaceans that are important to fisheries or ecological communities belong to the suborder Pleocyemata, the so-called "higher" Decapoda that includes crabs, lobsters, and shrimps. There is considerable incentive to study the embryonic and larval development of these arthropods because they are important to recruitment. However, within this suborder only stages in the latter two of the four ancestral larval phases of decapod development are free-living (mysis and decapodid), while all stages in the earlier two phases (nauplius and protozoea) have become "embryonized" and occur within the egg. Due to stage-specific

---

[*] Corresponding Author Email: bk.quinn@unb.ca.

variations in development rates, this makes understanding and predicting development and hatch times of such decapods very difficult, especially since the precise number and characteristics of embryonized stages are unknown. However, within the decapod suborder Dendrobranchiata, the so-called "lower" Decapoda that includes prawns in the superfamilies Penaeoidea and Sergestoidea, all larval stages in all four phases are free-living. This chapter examines whether it is possible to infer the types and durations of embryonized larval stages in higher decapod species based on equivalent free-living stages of dendrobranchiate prawns, assuming the dendrobranchiate development pattern is ancestral to that of the entire Decapoda. Stage-specific development times of dendrobranchiate crustaceans were obtained from previous studies and analyzed across species to quantify the proportions of total development spent in each stage and phase. Then, egg and larval development times of selected higher decapods were also collected from previous studies. Whether the proportion of total development spent by each of these species in their free-living larval stages relative to the total duration of all embryonized stages matched the equivalent proportion determined from the dendrobranchiate species examined was then tested. If these proportions matched, then dendrobranchiate-derived proportions could be used to estimate the durations of embryonized stages within higher decapods. This provides information that could be validated by later studies, for example by conducting egg dissections at predicted time periods to confirm the presence and types of different stages. If confirmed, this technique may allow for improved predictions of egg development rates and hatch times of commercially important higher decapods.

Keywords: larva, embryo, crustacean, Decapoda, Dendrobranchiata, Penaeoidea, Sergestoidea, Pleocyemata, development, embryonization

1. INTRODUCTION

The crustacean order Decapoda Latreille 1802 contains some of the most familiar species of this arthropod subphylum, including the crabs, hermit crabs, lobsters, prawns, and shrimps. Many decapod species support human fisheries of considerable economic importance (Anger 2001; Dall et al. 1990; MacKenzie 1988; Miller et al. 2016). Decapods also play important ecological roles in marine, freshwater, and terrestrial ecosystems as predators and prey, scavengers, and/or detritivores (Hamasaki et al. 2009;

Martin 2014a, b), as major components of secondary or tertiary ecosystem production (Corkett 1984; Corkett and McLaren 1970), and as invasive species (Hartnoll and Paul 1982; Nagaraj 1993). Some decapods (Reptantia Boas, 1880) spend most or all of their adult life crawling on the benthos or land surfaces, while others (Natantia Boas, 1880) swim regularly and may be found in planktonic, nektonic, or demersal marine or freshwater habitats as adults (Martin and Davis 2001).

Aside from a few groups of species that have evolved brooding or direct development, such as the freshwater and terrestrial crayfishes (suborders Astacoidea Latreille, 1802 and Parastacoidea Huxley, 1879 of the infraorder Astacidea Latreille, 1802) (Martin 2014a; Williamson 1982), most decapod species' life cycles contain one or more free-living larval phase. Most (but not all) decapod larvae are planktonic and/or pelagic, and they typically inhabit different environments than those used by adults of the same species (Anger 2001); this is especially true for benthic or terrestrial reptant decapods, but also applies to most natant species. Larvae are an essential part of all arthropod life cycles that contain them, which for some decapod crustacean species may be the main or only stage at which dispersal among habitats is possible (Anger 2001; Pechenik 1999; Pineda and Reyns 2018; Shanks 2009). The degree and nature of larval dispersal impacts decapods' population structures, range expansions, evolutionary adaptation potential, and even species persistence (Pechenik 1999). Larval supply to a decapod population is also an essential component of its recruitment (O'Connor et al. 2007; Phillips and Sastry 1980), and this is of particular interest to researchers working on fished decapod species (Miller 1997; Shumway et al. 1985). Potential recruitment of larvae is mediated by, among other things: (1) the duration and extent of dispersal (Reisser et al. 2013); and (2) the survival of larval throughout the dispersal process, which is usually thought to be inversely proportional to dispersal duration (Pineda and Reyns 2018; Shanks 2009).

The duration of larval development is strongly influenced by the temperatures experienced by larvae, with higher temperatures generally shortening the duration of development in a nonlinear manner (Anger 2001; MacKenzie 1988) so long as the species' tolerance limits are not exceeded

(Yamamoto et al. 2017). More rapid development is correlated with shorter dispersal and higher larval survival (O'Connor et al. 2007). Due to intra- and inter-seasonal changes in climate and oceanography or hydrography, larval dispersal is also influenced by exactly when larvae are released into the environment (Bradbury et al. 2000; Haarr 2018; Gendron and Ouellet 2009; Pineda and Reyns 2018). For example, a planktonic larva of a marine species could experience currents of differing strengths and directions, as well as different water temperatures, depending on what time of year it was released into the water column (Churchill et al. 2011; Haarr 2018; Pineda and Reyns 2018). A primary determinant of the timing of larval release, or hatch, is embryonic development. Similarly to larval development, embryonic development rates of decapods are inversely temperature-dependent (García-Guerrero and Hendrickx 2004, 2006a, b; Hamasaki et al. 2016). Therefore, the water temperatures experienced by embryos have an indirect impact on larval duration, dispersal, and likely survival due to their effect on hatch timing (Haarr 2018; Miller et al. 2016).

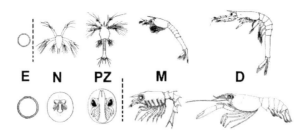

Figure 1. Examples of stages of potentially equivalent different developmental phases of decapod crustaceans in the suborders Dendrobranchiata (top) and Pleocyemata (bottom). Example species shown are *Penaeus monodon* (Dendrobranchiata: Penaeoidea: Penaeidae) and *Homarus americanus* (Pleocyemata: Astacidea). Developmental phases of Dendrobranchiata and their presumed equivalents among Pleocyemata are abbreviated as follows: E = embryo, N = nauplius ("egg-nauplius" and/or pre-pigmented embryo in Pleocyemata), PZ = protozoea (pigmented embryo in Pleocyemata), M = mysis, and D = decapodid (a.k.a. "postlarva", PL). Vertical dashed lines indicate the point at which hatch (i.e., eclosion from the egg membrane) occurs in each taxon. Figure made based on Silas et al. (1979) and Wahle et al. (2012).

There is thus much interest in quantifying development times of decapod embryos and larvae at different temperatures, as these have important implications to population dynamics, and thus to any fisheries or ecological processes that depend on a decapod species (Miller 1997; O'Connor et al. 2007). However, studying these processes, particularly embryonic development, is complicated by the fact that different developmental phases and stages occur within the life cycle of each species.

The larval phase of the life cycle is preceded by an embryonic phase, which is enclosed in an egg membrane and has no obvious appendages or larval morphology aside from limb buds and anlages later in the phase (García-Guerrero and Hendrickx 2004, 2006a, b; Talbot and Helluy 1995). The larval phase can itself be divided up into different phases, which are made up of morphologically distinct larval types. Among the Decapoda, four larval phases are generally recognized (Williamson 1982; Martin 2014b; Martin et al. 2014b; see also Figure 1), which in order of increasing complexity, size, and time elapsed after the embryonic phase are: (1) the nauplius phase, a "head larva" in which cephalic limbs (e.g., antennae) are used for locomotion and functional lateral compound eyes and thoracic appendages are not present; (2) the protozoea phase, in which lateral compound eyes, thoracic appendages (pereiopods), and the abdomen progressively develop and become more prominent, but cephalic appendages are still used for locomotion; (3) the mysis phase, in which the function of locomotion switches to the exopodites of the thoracic appendages and abdominal appendages (pleopods and uropods) progressively develop, although they remain largely non-functional; and (4) the decapodid phase (also often referred to as the "postlarva" or megalopa phase, among other names; Felder et al. 1985), in which morphology becomes very similar to that of juveniles and adults, the exopodites of the thoracic appendages are typically lost, locomotory function switches to the abdominal appendages (pleopods), and migrations and/or settlement to the juvenile habitat (often the benthos) followed by metamorphosis to the juvenile phase occur (Oliphant et al. 2013; Martin 2014b; Martin et al. 2014b). Some researchers have alternatively combined the protozoea and mysis phases into a single phase, the zoea phase (Martin et al. 2014b), meaning that there are actually three phases. However, since most studies of dendrobranchiates still separate the zoeal stages into protozoea and mysis stages, the four-phase terminology is used within this chapter.

Each larval phase can contain one or more distinct stages, which generally resemble the morphology of the overall phase but differ slightly from one another in certain aspects; for example, early stages within a phase often lack appendages that appear and develop in later stages within the same phase (Williamson 1982). Each larval stage (and phase) is separated from the next one by a moult (Martin et al. 2014b). In some species, moults can occur that allow the larva to grow without any developmental progression occurring, resulting in there being multiple moults within the same stage; in this case, the different "sub-stages" are referred to as instars (Goldstein et al. 2008; Martin et al. 2014b; Oliphant et al. 2013). Larvae of most species must complete development through all of their phases, and typically the number and types of stages are also highly consistent within a species (Gore 1985; Williamson 1982); it should be noted, though, that under some circumstances many species can alter their developmental pathways to skip or add stages, for example at stressful extreme temperatures, with low food supplies, or if settlement habitat cannot be found (Gore 1985; Oliphant et al. 2013; Quinn 2016).

Importantly, in most (if not all) species the duration of each developmental stage (and phase) for the same species at the same temperature are typically not the same (i.e., development is not isochronal; Quinn 2018); for example, generally (but not always) developmentally later stages require more time to be completed than earlier ones (Gore 1985; MacKenzie 1988; Williamson 1982). Further, the exact quantitative relationship between temperature and development time can vary among developmental stages (and therefore phases) (Anger 2001; MacKenzie 1988). As a result, there is potential for considerable errors to be made if one attempts to predict total development times across multiple stages or phases without accounting for stage-specific development times. This is especially important for predicting development times in variable thermal regimes, such as those inhabited by decapod embryos and larvae in nature, as assuming one constant temperature-development relation across multiple stages or phases will miss variations and accumulate errors in the development times of different individual stages.

Thus, to assess the temperature-dependent development of decapod larvae, newly-hatched individuals are typically reared in laboratory settings at different, controlled temperatures (e.g., Goldstein et al. 2008; Hamasaki et al. 2009; MacKenzie 1988; Nagaraj 1993). When moults or morphological changes indicating that a moult occurred are observed, this allows individual larval stages (and the phases that consist of them) to be delineated, and their temperature-dependent durations quantified. For embryonic development, however, there is an added challenge in that all developmental stages are completed within the egg membrane (Figure 1), making them effectively unobservable in live specimens with current methodologies. This is not much of a problem for the study of prawns in the decapod suborder Dendrobranchiata Spence Bate 1888, because their embryonic phase is typically very short (i.e., often completed within a few hours) due to the fact that their planktonic eggs are not brooded and quickly hatch into free-living larvae (Dall et al. 1990; Martin et al. 2014a; Taveras and Martin 2010). However, for diverse members of the decapod suborder Pleocyemata Burkenroad 1963 (crabs, lobsters, shrimps, etc.), the embryos are brooded by females for extended periods lasting up to several months (or even exceeding a year) and undergo several changes, perhaps corresponding to different stages or phases, during this time (García-Guerrero and Hendrickx 2004, 2006a, b; Talbot and Helluy 1995); therefore assessing their embryonic stages may be of considerable importance to predicting their development and hatch times.

Different points in the embryonic development of species of Pleocyemata can be sampled through the destructive dissection of the egg, and this has been done to identify different approximate "stages" during the development of some species, including for example different cell-numbered stages of early embryos, the "egg-nauplius", and a more developed egg-nauplius termed a "metanauplius" (Alwes and Scholtz 2006; García-Guerrero and Hendrickx 2004, 2006a, b; Talbot and Helluy 1995). Some workers, through careful observations, have also observed evidence of possible setal changes and embryonic moults occurring within the egg (e.g., Kidd 1991; Talbot and Helluy 1995), but these reports are not confirmed or accepted by all other researchers. In the embryos of pleocyematan decapods, eventually a pigmented pair of compound eyes can be recognized, and it has been shown that the size of these eyes increases linearly or logarithmically as development progresses at a rate that is temperature-dependent (Perkins 1972; Gendron and Ouellet 2009; Miller et al. 2016). Before the onset of pigmentation, there is no established means to measure or estimate embryonic development rate (other than perhaps prior experience), and there is no consensus regarding whether and how many moults or different embryonic stages occur during this period, let alone how long they are at different temperatures. After an embryo's eyes become pigmented, its developmental progression can be estimated based on changing eye size (e.g., the Perkins (1972) Eye Index), or other indirect indices such as decreasing yolk content (Giovagnoli et al. 2014). However, these indirect indices tend to be highly variable among embryos, and the accuracy with which they measure actual developmental progression is uncertain; for instance, these measure the size of embryonic features and assume continuous growth, whereas size is usually a very coarse indicator of development at best (Forster et al. 2011) and mainly increases in a stepwise fashion, due to the requirements of crustacean larvae (likely even when embryonized) to moult in order to increase in size and proceed in development (Anger 2001). This approach also ignores the potential occurrence of different developmental stages within the pigmented embryonic phase. As a result, studies attempting to predict the timing of hatch using these methods for commercially important species, such as the American lobster *Homarus americanus* H. Milne Edwards 1837 (Pleocyemata: infraorder Astacidea: family Nehpropidae Dana 1852) for example, frequently produce erroneous estimates or encounter unexplainable variation in predicted hatch timing (Gendron and Ouellet 2009; Haarr 2018; Miller et al. 2016).

It could be far more accurate to determine which embryonized larval stages occur during the embryonic development of a given species (i.e., how many different phases and stages there are) and how much time is required to complete them at different temperatures. Later studies could then use these stage-specific development curves to predict the time required for embryonic development to be completed in different thermal regimes. While this cannot be done directly at present, inferences may be made based on development in the Dendrobranchiata applied to the Pleocyemata, since some or all of the stages occurring in the development of embryos of the Pleocyemata may be homologous to larval stages that are still free-living in the Dendrobranchiata.

Crustaceans belonging to the superfamilies Penaeoidea Rafinesque 1815 and Sergestoidea Dana 1852 (both in the suborder Dendrobranchiata, also known as the "lower" Decapoda) are the only members of the Decapoda to possess four free-living larval phases (Figure 1), consisting of the nauplius, protozoea, mysis, and decapodid stages (Martin 2014b; Martin et al. 2014b; Taveras and Martin 2010; Williamson 1982). In all other ("higher") decapods (Pleocyemata), only the mysis and decapodid stages (if even these) are free-living, while the nauplius and protozoea stages have become "embryonized" and are presumed to be completed entirely within the egg (Bressan and Müller 1997; Jirkowski et al. 2013; Martin 2014b; Müller et al. 2004; Williamson 1982; see also Figure 1). As a consequence of this embryonization, precise delineation of specific phases and stages within the development of embryos in the higher Decapoda has not been possible. However, it may be possible to apply observations of development of all four free-living larval phases and their constituent stages in the Dendrobranchiata to infer probable numbers and relative durations of embryonized larval stages in higher decapods. The Dendrobranchiata and Pleocyemata diverged from a common ancestor, and based on development in other groups of the class Malacostraca Latreille, 1802 (to which the Decapoda belong) that have a free-living nauplius phase (e.g., the order Euphausiacea Dana, 1852, which includes krill; Akther et al. 2015; Martin and Gómez-Gutiérrez 2014; Ross 1981) this ancestor is presumed to have had a free-living nauplius (and likely also protozoea) phase in its development (Martin 2014a; Martin et al. 2014c). Thus, the development pattern of dendrobranchiates may be considered to be ancestral to that of the Pleocyemata.

Most members of the Dendrobranchiata are marine prawns, and after a brief planktonic embryonic phase they develop through 2 (Seregestoidea) or 5-6 and rarely 8 (Penaeoidea) nauplius, 3 protozoea, 2 (sergestoids) or 2-5 (penaeoids) mysis, and a few to many (i.e., 1-9) decapodid stages (Dall et al. 1990; Martin et al. 2014a; Taveras and Martin 2010; Williamson 1982). Members of the Pleocyemata, on the other hand, develop through 1-15 (but usually 2-8 and often ca. 3) mysis stages and 1 or more (perhaps up to 12) decapodid stages, depending on species (Williamson 1982), meaning that any nauplius and protozoea stages they have are completed within the egg. Embryos contained within the eggs of species of Pleocyemata do appear to go through a nauplius-like phase ("egg-nauplius"; Jirkowski et al. 2013), and the acquisition of pigments by the eyes of embryos may be taken as being equivalent to the transition to the protozoea phase of development, as dendrobranchiate nauplii lack large, pigmented lateral compound eyes but their protozoeae have them (Martin et al. 2014a). There may also be distinct stages within each of these embryonic phases, which are responsible for some of the observed variability in predicted hatch times.

It is therefore conceivable that if the proportion of total development spent by dendrobranchiates within each larval phase and stage is consistent among species in the superfamily Dendrobranchiata and their development is homologous to that of the Pleocyemata (Figure 1), then the proportion of development spent within the mysis (and perhaps decapodid) stages could be applied to the equivalent stages of pleocyematan species. One could then conceivably "hindcast" the probable time of occurrence and duration of the embryonized stages at the same temperature based on the proportion of development spent in these stages by dendrobranchiates. This could then produce stage-specific development curves for all embryonic stages and/or phases, which could then be used to predict egg development times in different thermal regimes. These inferred stages and their developmental curves could then be used to predict time of hatch with higher potential accuracy than is currently possible. This chapter reports on a first attempt to do this using data obtained from literature searches and subjected to preliminary meta-analyses. Observed development times of pleocyematan embryos were compared against predictions of these made based on dendrobranchiate development, in an attempt to validate this method. This work creates a foundation on which future work could be built, for example to confirm whether inferred stages actually occur, and whether embryonic stages exhibit plasticity (e.g., under different conditions, the number and types of stages passed through also changes).

2. MATERIALS AND METHODS

2.1. Obtaining Dendrobranchiate Development Time Data

The decapod suborder Dendrobranchiata is divided into the superfamilies Penaeoidea and Sergestoidea. Currently, the Penaeoidea are further subdivided into the families Aristeidae Wood-Mason, 1891, Benthesicymidae Wood-Mason, 1891, Penaeidae Rafinesque, 1815, Sicyoniidae Ortmann, 1898, and Solenoceridae Wood-Mason, 1891, while the Sergestoidea are subdivided into the families Luciferidae De Haan, 1849 and Sergestidae Dana, 1852 (Martin and Davis 2001; Martin et al. 2014a; WoRMS Editorial Board 2018).

To obtain information on stage- and/or phase-specific development times of dendrobranchiate crustacean larvae, a series of online literature searches were conducted in March-May 2018 through Web of Science (Clarivate 2018) and Google Scholar (Google 2018). Search terms used included various combinations of "dendrobranchiat*", "penae*", "sergest*", "development", and "larva", as well as common dendro-branchiate genus names (e.g., the type genera of the various families in this suborder: *Aristeus* Duvernoy, 1840, *Benthesicymus* Spence Bate, 1881, *Penaeus* Fabricius, 1798, *Sicyonia* H. Milne Edwards, 1830, *Solenocera* Lucas, 1849, *Sergestes* H. Milne Edwards, 1830, and *Lucifer* J.V. Thompson, 1829) obtained from WoRMS, the World Register of Marine Species (WoRMS Editorial Board 2018). Cited reference searches were also conducted on any relevant papers obtained from these searches, and also on several reviews written about the Dendrobranchiata or its subgroups (Dall et al. 1990; Martin et al. 2014a; Taveras and Martin 2010).

To provide data useful for the present analyses, a paper had to report the developmental durations (in hours (h) or days (d)) of multiple larval phases (embryo, nauplius, protozoea, mysis, and/or decapodid) of a dendrobranchiate species reared under the same conditions, and optimally would also report the duration of individual larval stages within these phases. This requirement ended up being surprisingly restrictive, as the majority of studies obtained from literature searches (that the author could access) reported on the morphology, identification, feeding regimens, and/or survival of larvae under different rearing conditions, but did not report development times of specific phases or stages. As a result, useful data could only be obtained from 15 studies for a total of 13 different dendrobranchiate species (Table 1). Almost all (11) of the species for which data could be obtained belonged to the family Penaeidae within the superfamily Penaeoidea, while the remaining 2 species belonged to the family Sergestidae within the superfamily Sergestoidea (Table 1). Thus, unfortunately, data for the remaining 4 penaeoid families and 1 sergestoid family, and several genera within all dendrobranchiate families, were not included in the present chapter's analyses.

Data were extracted from the studies obtained in literature searches, either directly from development times reported in the text or tables of these papers, or by using the program ImageJ (Schneider et al. 2012) to estimate values presented in their figures. Most studies (Table 1) only reported a single value (i.e., a mean or median) or range for the duration of each phase or stage, which simplified data extraction but limited the types of analyses that could be performed on the data obtained

Table 1. Species of prawns belonging to the decapod suborder Dendrobranchiata

Species	Temperature (°C)	Other treatments	Stages reported[a,b]
Superfamily Sergestoidea Dana, 1852: Family Sergestidae Dana, 1852			
Eusergestes similis (Hansen, 1903)[1]	10 and 14		E, N1-N4, PZ1-PZ3, M1-M2, PL1-PL7[c]
Lucenososergia lucens (Hansen, 1922)[2]	20		E, N1-N2, PZ1-PZ3, M1-M2, PL1-PL5
Superfamily Penaeoidea Rafinesque, 1815: Family Penaeidae Rafinesque, 1815			
Metapenaeus affinis (H. Milne Edwards, 1837)[3]	29.2-30.6	2 replicate experiments	E, N1-N6, PZ1-PZ3, M1-M3, PL1-PL6
Metapenaeus brevicornis (H. Milne Edwards, 1837)[4]	29.2-30.8		E, N1-N6, PZ1-PZ3

Species	Temperature (°C)	Other treatments	Stages reported[a,b]
Metapenaeus dalli Racek, 1957[5]	26		E, N[1-6], PZ1-PZ3, M1-M3, PL1
Metapenaeus ensis (De Haan, 1844)[6]	26-28		E, N[1-4], PZ1-PZ3, M1-M3
Metapenaeus monoceros (Fabricius, 1798)[7]	28	Salinities: 25[d], 30, 35, 40, 45, 50, and 55[d] ppt	PZ[1-3], M[1-3]
Penaeus californiensis Holmes, 1900[8,9]	28-30[8]		E, N1-N6, PZ1-PZ3, M1-M3
	22, 25, 28, and 30[9]		N[1-5], N6, PZ1-PZ3, M1-M3, PL1
Penaeus duorarum Burkenroad, 1939[10]	21, 26	Two development pathways (3 or 4 M stages)	N[1-6], PZ1-PZ3, M1-M3/M4
Penaeus latisulcatus Kishinouye, 1896[11,12]	17, 20, 22.5, and 25[11]		N[1-6], PZ[1-3], M[1-3]
	29.23[12]		N[1-6], PZ[1-3], M[1-3]
Penaeus monodon Fabricius, 1798[13]	26.5-28.5		E, N1-N6, PZ1-PZ3, M1-M3, M4? (M3-PL intermediate)
Penaeus semisulcatus De Haan, 1844[14]	22[c], 26, 30, and 34	Two experiments; salinities (Expt. 2): 25, 30, and 35 ppt	PZ[1-3], M[1-3]
Trachysalambria curvirostris (Stimpson, 1860)[15]	25.8-26.2		E, N1-N6, PZ1-PZ3, M1-M3

Notes: [a]E = embryo, N = nauplius, PZ = protozoea, M = mysis, and PL = "postlarva" (i.e., decapodid); [b]If stage durations were reported for each individual stage in a phase, the notation N1-Nn, PZ1-PZn, or M1-Mn is used (where n = total # stages in phase), whereas if only the total phase duration is reported the notation N[1-n], PZ[1-n], or M[1-n] is used; [c]PL stages were not considered in this chapter; [d]Only the PZ phase, and not the M phase, was completed at 25 and 55 ppt; [e]22°C only tested in Expt. 1, not in Expt. 2. Development time data was obtained from literature searches for analyses in this chapter. Details of the rearing temperatures and other experimental treatments in original studies, as well as for which stages and/or phases data were available, are also summarized

Sources of data for each species were: [1]Omori (1979); [2]Omori (1971); [3]Thomas et al. (1974); [4] Rao (1979); [5]Crisp et al. (2016); [6]Leong et al. (1992); [7]Kumlu et al. (2001); [8]Kitani and Alvarado (1982); [9]Villarreal & Hernandez-Llamas (2005); [10]Ewald (1965); [11]Roberts et al. (2012); [12]Shokita (1984) [data cited by Roberts et al. (2012)]; [13]Silas et al. (1979); [14]Kumlu et al. (2000); [15]Ronquillo and Saisho (1995).

When a range was reported, the midpoint was used as the duration for analyses. If development times were reported under multiple experimental conditions, for example different rearing temperatures and/or salinities, an average value was calculated for each species and stage or phase across all reported conditions (see Table 1). If total phase durations were reported, these were extracted directly from the source paper, whereas if they were not reported they were calculated by summing the average durations of all stages within a phase.

Many studies reported development times for one or more (up to 7) stage(s) within the "postlarva" (PL) phase, more properly called the decapodid phase (D in Figure 1; Martin et al. 2014b), or reported intermediate-stage larvae between the last mysis and first PL stages. For the purposes of this chapter, however, development times of these stages were not included in any calculations or analyses. These were omitted because: (1) the morphology of dendrobranchiates changes very gradually between PL stages, such that PL stages gradually grade from decapodid morphology into juvenile morphology (Martin et al. 2014a; Williamson 1982), and thus there is no clear definition of exactly how many PL stages any species has and specifically when the PL phase ends and the juvenile phase begins; and (2) in both dendrobranchiate and pleocyematan decapods the decapodid (or PL) phase is typically the settling stage of the life cycle, and its length and/or the number of stages within the phase can vary considerably as a function of developmental or settlement delay (Oliphant et al. 2013; Quinn 2016 and references therein). It was thus considered likely that including stages in the decapodid phase would introduce extra variability and noise into the dataset, leading to potential errors, so these were omitted.

In this chapter, then, the total developmental duration of a dendrobranchiate species was defined as the time elapsed between the extrusion of eggs to the moult from the last mysis to the first decapodid (PL) stage; this time period thus included the entirety of the pre-naupliar embryonic (E), nauplius (N), protozoea (PZ), and mysis (M) phases, and all stages therein. Total development time was calculated by summing the mean durations of all developmental phases for a given species. It should be noted that not all studies obtained reported durations of all developmental phases (see Table 1), so this total development time value and associated developmental ratios and proportions could not be calculated for these species; however, ratios of the durations of the developmental phases reported in such studies could still be used in some calculations and analyses, as outlined in the following sections.

2.2. Calculation of Developmental Ratios and Proportions for Dendrobranchiate Phases

For each of the 13 dendrobranchiate species for which data were obtained, the average duration (in d) of each developmental phase (E, N, PZ, and M) was divided by that of each other phase to calculate developmental ratios; ratios were also calculated between certain individual phases and combinations of other phases. Of particular importance were the ratios between the M phase and the combined [E + N + PZ] and [E + N] phases, as these were equivalent to the total embryonic and pre-pigmented embryonic phases of the Pleocyemata, respectively. Developmental proportions for each species and phase were also calculated by dividing the average phase duration by the total developmental duration (phases [E + N + PZ + M]) if available, and also by the combined duration of the [N + PZ + M] phases because not all studies reported durations of the relatively short E phase (Table 1). The number of stages in each of the N, PZ, and M phases of each species was also noted.

Mean values ± standard error (SE) of all of the above developmental characteristics (ratios, proportions, and numbers of stages) were then calculated across (1) all 11 penaeid species, (2) both sergestid species, and (3) all 13 dendrobranchiate species. Several of these values were then used in subsequent calculations and analyses. Differences in phase ratios and proportions were discussed qualitatively, and later also roughly compared with equivalent characteristics of developmental phases in the Pleocyemata. Because replicate values were not available for most species, formal statistical comparisons of these characteristics among dendrobranchiate species (i.e., to test whether they were consistent across species) were not possible; the amount of interspecific variation was therefore examined and discussed qualitatively, for example by considering the size of error measures (SE) about the mean.

2.3. Calculation and Comparisons of Developmental Proportions for Dendrobranchiate Stages

The proportion of total development spent by each dendrobranchiate species in each developmental stage was also calculated similarly to the proportions for phases described above. Stage-specific proportions were initially calculated by dividing stage durations by the total duration of development [E + N + PZ + M] for each species, but because E durations were not always reported this resulted in relatively low sample sizes and perhaps poorer estimates. Proportions were also alternatively calculated by dividing stage durations by the combined duration of the [N + PZ + M] phases. Because the E stage was always relatively short, its omission had minimal impacts on the proportions calculated, and in fact statistical comparisons made on both sets of proportion data had nearly identical results. Therefore, only results for the latter set of proportions (without the E phase included in the denominator) are reported herein.

These stage-specific proportions were then compared among different developmental stages and between larger taxonomic groups (penaeids and sergestids) using a two-way analysis of variance (ANOVA). The square-roots of proportion data were arcsine-transformed before performing the ANOVA to meet the assumptions (normality of residuals and homogeneity of variances) of parametric tests. Due to a statistically significant interaction between stage and group (see Results), separate one-way ANOVAs comparing proportions among stages were carried out for penaeids and sergestids; for the sergestids, only the first two nauplius stages (N1 and N2) were included in this analysis, as the third and fourth N stages were only reported for one species. A one-way ANOVA ignoring group to compare proportions among the stages of all dendrobranchiate species combined was also performed. Tukey's honestly significant difference (HSD) test was used to perform post-hoc comparisons among stages if a significant effect of stage on proportion was found. All statistical analyses were performed in IBM SPSS Statistics 23 (SPSS Inc., 2015).

2.4. Obtaining Development Time Data for Species in the Pleocyemata

The decapod suborder Pleocyemata currently consists of 11 accepted infraorders (WoRMS Editorial Board 2018; Poore 2016). To assess whether developmental ratios and/or proportions calculated for free-living dendrobranchiate larvae could be used to predict development times of the equivalent embryonized stages of the Pleocyemata, data from one representative species of each pleocyematan infraorder were sought out by performing further literature searches if needed. The aim was to find a species from each infraorder for which one or more studies had reported development times of embryos (i.e., eggs) from extrusion to hatch, and under overlapping conditions (e.g., rearing temperatures) development times of all larval (mysis) stages were also reported.

Optimally, studies were sought that also subdivided egg development into pre-pigmented and pigmented stages, which were herein treated as being equivalent to the dendrobranchiate [E + N] and PZ phases, respectively, although this was not always attainable; reporting of individual mysis stage durations was also preferred. As for the dendrobranchiates, decapodid or "postlarva" (megalopa, glaucothoea, puerulus, etc.) stages were not considered. For several infraorders, the author was already in possession of larval development time data obtained for another meta-analysis study (Quinn 2018), so to simplify the searching process if embryonic development time data under equivalent conditions could be found for these species they were used in the present analyses.

Detailed developmental data could not be found for both embryos and larvae of any representative members of the infraorders Glypheidea Van Straelen, 1925, Polychelida Scholtz & Richter, 1995, and Procarididea Felgenhauer & Abele, 1983, so these had to be excluded from further analyses. Of the remaining 8 infraorders, durations of embryonic and larval development at overlapping temperatures were obtained for one species each from 16 different source studies (Table 2), although embryonic sub-stage

Brady K. Quinn

Table 2. Species representing 8 of the 11 currently accepted infraorders of the decapod suborder Pleocyemata

Infraorder	Species	Temperature overlap range (°C)	Proportion of total development in each phase(s) (range)[a]			
			E + N (0.19)[b]	PZ (0.47)[b]	E + N + PZ (0.66)[b]	M (0.34)[b]
Achelata Scholtz & Richter, 1995[1,2]	*Panulirus argus* (Latreille, 1804)	26-27	0.024	0.043	0.067	0.933
Anomura MacLeay, 1838[3,4]	*Birgus latro* (Linnaeus, 1767)	24-29	0.291-0.295	0.235-0.288	0.526-0.583	0.417-0.474
Astacidea Latreille, 1802[5,6]	*Homarus americanus* H. Milne Edwards, 1837	10-22	0.205-0.217	0.654-0.690	0.859-0.908	0.092-0.141
Axiidae de Saint Laurent, 1979[7]	*Nihonotrypaea japonica* Ortmann, 1891	20.5-24.5	0.313-0.355	0.130-0.133	0.444-0.488	0.512-0.556
Brachyura Latreille, 1802[8,9]	*Carcinus maenas* (Linnaeus, 1758)	11-25	n.d.	n.d.	0.376-0.628[c]	0.372-0.624[c]
Caridea Dana, 1852[10,11,12]	*Pandalus borealis* Krøyer, 1838	2-5	0.404-0.440	0.309-0.312	0.716-0.749	0.251-0.284
Gebiidae de Saint Laurent, 1979[13,14]	*Upogebia africana* (Ortmann, 1894)	17.5-24	n.d.	n.d.	0.792-0.805	0.195-0.208
Stenopodidae Spence Bate, 1888[15,16]	*Stenopus hispidus* (Olivier, 1811)	26-27	n.d.	n.d.	0.156	0.844

Notes: "n.d." means no data were available for this specific phase; [a] E = embryo, N = nauplius, PZ = protozoea, and M = mysis (in this table, these are equivalent phases of Pleocyemata, see Figure 1); [b]Numbers in parentheses are average proportions for all dendrobranchiates (see Table 3); [c]For this species, the proportions of development were [E + N + PZ] ≥ M at temperatures ≤ 15°C, but [E + N + PZ] < M at temperatures > 15°C.

Embryonic and larval development time data were obtained from published studies and used in analyses of the predictive abilities of dendrobranchiate developmental proportions (Table 3) in this chapter. The range of temperatures over which the embryonic and larval development time data obtained overlapped is shown. The mean actual proportion of total development spent in each phase ([E + N] = pre-pigmented embryo, PZ = pigmented embryo, M = larvae) was calculated for each species, at each temperature; values shown are the range of mean proportions across all temperatures

Sources of data for each species were: [1]Embryos: Ziegler and Forward, Jr. (2007); [2]Larvae: Goldstein et al. (2008); [3]Embryos: Hamaski et al. (2016); [4]Larvae: Hamasaki et al. (2009); [5]Embryos: Perkins (1972); [6]Larvae: MacKenzie (1988); [7]Embryos and larvae: Tamaki et al. (1996); [8]Embryos: Hartnoll and Paul (1982); [9]Larvae: Nagaraj (1993); [10]Embryos (total duration): Brillon et al. (2005); [11]Embryos (pigmented phase only): Stickney (1982) [cited in Shumway et al. (1985)]; [12]Larvae: Ouellet and Chabot (2005); [13]Embryos: Hill (1977); [14]Larvae: Newman et al. (2006); [15]Embryos: Gregati et al. (2010); [16]Larvae: Fletcher et al. (1995).

durations (to distinguish [E + N] vs. PZ equivalents) and/or data at multiple temperatures could not be obtained for all species. Data were extracted from studies and total developmental durations (embryonic + larval) were calculated as described above for the Dendrobranchiata. To simplify calculations (and because some studies only reported single values for the durations of certain developmental phases), a single mean duration for each stage, rearing temperature, and species was calculated. Importantly, although all studies obtained for the same species reported durations over overlapping temperature ranges, the temperatures for which durations were reported were not always exactly the same. Therefore, to achieve matching embryonic-larval development time sets at each temperature, power functions (development time = $a*temperature^b$) were fit to mean development times of each phase of pleocyematan development at different temperatures, and used to estimate development times at other temperatures.

2.5. Calculations and Comparisons of Developmental Proportions for Pleocyematan Phases

The mean ± SE observed proportion of the total developmental period spent by each species of Pleocyemata examined in the combined [E + N + PZ] (i.e., embryonic) phase and the total M (i.e., larval) phases was calculated across all rearing temperatures and conditions; when possible, the mean proportions of development spent in the combined [E + N] (i.e., pre-pigmented egg) and PZ (i.e., pigmented egg) phases were also calculated. These proportions were examined for each species, and qualitatively compared among pleocyematan infraorders, as well as to the equivalent proportions calculated earlier for the Dendrobranchiata. This was done to examine any taxonomic trends in developmental programs among the Pleocyemata, and also to assess the likelihood that dendrobranchiate-derived proportions could be used in combination with pleocyematan M phase durations to predict embryonic durations of the Pleocyemata.

Brady K. Quinn

2.6. Prediction of Pleocyematan Phase and Stage Durations and Comparisons with Observations

In principle, if dendrobranchiate-derived developmental proportions provide a reasonable approximation of equivalent proportions of the stages of development in species of Pleocyemata, then the average proportions of total development spent in each stage calculated across all Dendrobranchiata (or perhaps for penaeids and/or sergestids separately) could be used to estimate when during embryonic development different embryonized larval stages should be expected to occur at a given temperature. This would be done by dividing the proportion of dendrobranchiate development for a given stage by the proportion of dendrobranchiate development spent in the M phase, and then multiplying the resulting ratio by the observed M duration of a pleocyematan decapod at a given temperature. As an example, this was done for the American lobster, *H. americanus* (Astacidea), using proportions calculated for penaeids, sergestids, and all dendrobranchiates. Data are not currently available to assess the actual timing of such stages, however, so these stage predictions cannot be used to assess the accuracy of this method; rather, estimates of phase durations must be used for this.

For each of the 8 pleocyematan species, the duration of the total embryonic period [E + N + PZ] at each tested temperature was predicted by multiplying the observed M duration of that species at that temperature by the average ratio of the M phase to the [E + N + PZ] phases that was calculated for penaeids, sergestids, or all dendrobranchiates (Table 3). The duration of the pre-pigmented [E + N] and pigmented (PZ) embryonic stages could also be estimated based on observed M duration and the ratios of M/[E + N] and M/PZ, respectively, calculated herein for different groups of dendrobranchiates, although this was only done for species for which actual observed durations of these embryonic sub-phases were available.

Predicted phase durations were then compared to actual observed embryonic durations for each species across all temperatures. For two

Table 3. Mean ± SE values of the numbers of stages, ratios of durations between phases, and proportions of total development with or without the short embryonic phase included spent in each phase calculated across all penaeid, sergestid, and dendrobranchiate species examined

Developmental characteristic	Penaeidae	Sergestidae	All Dendrobranchiata
# of stages in phase[a]			
N	5.780 ± 0.222 (9)	3.0 ± 1.0 (2)	5.270 ± 0.407 (11)
PZ	3.0 ± 0.0 (11)	3.0 ± 0.0 (2)	3.0 ± 0.0 (13)
M	3.10 ± 0.067 (10)	2.0 ± 0.0 (2)	2.920 ± 0.469 (12)
Ratios of phase durations			
E/N	0.36 ± 0.069 (6)	0.39 ± 0.029 (2)	0.37 ± 0.051 (8)
E/PZ	0.13 ± 0.022 (6)	0.08 ± 0.013 (2)	0.11 ± 0.018 (8)
E/M	0.14 ± 0.023 (5)	0.23 ± 0.006 (2)	0.17 ± 0.023 (7)
N/PZ	0.37 ± 0.024 (9)	0.21 ± 0.051 (2)	0.34 ± 0.028 (11)
N/M	0.42 ± 0.043 (8)	0.60 ± 0.029 (2)	0.45 ± 0.042 (10)
PZ/M	1.02 ± 0.150 (10)	3.02 ± 0.602 (2)	1.35 ± 0.267 (12)
[N + PZ]/E	12.45 ± 1.993 (6)	15.77 ± 1.969 (2)	13.28 ± 1.60 (8)
[N + PZ]/M	1.59 ± 0.173 (8)	3.61 ± 0.574 (2)	1.99 ± 0.315 (10)
E/[N + PZ + M]	0.06 ± 0.008 (5)	0.05 ± 0.005 (2)	0.06 ± 0.006 (7)
[E + N]/M	0.53 ± 0.071 (5)	0.83 ± 0.027 (2)	0.62 ± 0.074 (7)
[E + N + PZ]/M	1.65 ± 0.270 (5)	3.84 ± 0.580 (2)	2.28 ± 0.463 (7)
Proportions of total development from E to PL [E + N + PZ + M] in phase			
E	0.05 ± 0.007 (5)	0.05 ± 0.004 (2)	0.05 ± 0.005 (7)
N	0.14 ± 0.009 (5)	0.13 ± 0.021 (2)	0.14 ± 0.008 (7)
PZ	0.41 ± 0.029 (5)	0.62 ± 0.051 (2)	0.47 ± 0.045 (7)
M	0.39 ± 0.035 (5)	0.21 ± 0.025 (2)	0.34 ± 0.042 (7)
[N + P + Z]	0.55 ± 0.036 (5)	0.74 ± 0.030 (2)	0.61 ± 0.043 (7)
Proportions of total larval development from N to PL [N + PZ + M] in phase			
N	0.16 ± 0.010 (8)	0.13 ± 0.023 (2)	0.15 ± 0.009 (10)
PZ	0.44 ± 0.022 (8)	0.65 ± 0.050 (2)	0.48 ± 0.033 (10)
M	0.40 ± 0.026 (8)	0.22 ± 0.027 (2)	0.36 ± 0.032 (10)

Note: [a] E = embryo, N = nauplius, PZ = protozoea, and M = mysis.

Numbers in parentheses are the number of species for which there was enough data to calculate each mean value

species of Pleocyemata [*Panulirus argus* (Latreille, 1804) (Achelata Scholtz & Richter, 1995) and *Stenopus hispidus* (Olivier, 1811) (Stenopodidae

Spence Bate, 1888)] only one data point (temperature) was available, so comparisons for these species were limited to describing whether and by how much predictions over- or underestimated observed embryonic durations. For all other species, paired t-tests were conducted to test whether observed and prediction durations were significantly different, with different rearing temperatures serving as replicates in these tests. Separate tests were performed for each phase ([E + N + PZ], [E + N], and PZ), dendrobranchiate group (penaeids, sergestids, and all dendrobranchiates), and species. The null hypothesis in all these tests was that if dendrobranchiate-derived proportions provide a good representation of pleocyematan development, then the difference between predicted and observed phase durations should not be significantly different from zero.

Finally, to further assess whether and how predicted and observed embryonic phase durations differed, absolute differences and absolute percent (%) differences (relative to observed values) between all predicted and observed durations were calculated, and mean ± SE values of these difference were calculated across all species for each phase and dendrobranchiate group. Overall paired *t*-tests were also conducted across all species to compare observed and predicted embryonic phase durations for each phase and dendrobranchiate group.

3. RESULTS

3.1. Proportions of Dendrobranchiate Development Spent in Different Phases

The average ± SE relative ratios of the durations of different developmental phases or combinations of phases of the 13 dendrobranchiate species for which data were obtained, as well as the proportion of development encompassed by each phase, are presented in Table 3. Across all dendrobranchiates, the embryonic (E) phase was the shortest, taking up approximately 5% of the total developmental period, followed by the nauplius (N) phase, which took up 14% of total development (Table 3). The majority of the developmental period (47%) was spent in the protozoea (PZ) phase, followed by the mysis phase (M), which took up 34% of development (Table 3). Because the majority of species (11/13) examined were penaeids, the average proportions and ratios for each phase were similar between those calculated for the Penaeidae only versus those for all dendrobranchiates examined (Table 3). Among the 2 sergestid species examined, the proportions of development spent in the E and N phases were similar to the penaeid and dendrobranchiate averages, but the PZ phase was comparatively longer (62%) and the M phase comparatively shorter (21%) (Table 3). In general, there was little variation in phase duration ratios or the proportion of development spent in each phase among dendrobranchiate species (with the exception of differences for the PZ and M phases between penaeids and sergestids just mentioned; Table 3). This low variation is evident from the relatively small SEs of mean proportions for each phase out of total development, which only ranged from 0.4 to 5.1% of the developmental period (Table 3). The number of stages within each developmental phase barely varied at all among penaeids, with there almost always being 6 nauplius, 3 protozoea, and 3 mysis stages (Table 3). Among the sergestids considered, there were 2 or 4 nauplius stages, 3 protozeae, and 2 mysis stages (Table 3).

3.2. Proportions of Dendrobranchiate Development Spent in Different Stages

The relative proportion of development spent by dendrobranchiates in each stage was calculated in terms of the proportion of the total duration of the [N + PZ + M] phases because not all studies reported durations of the E phase (Figure 2). These proportions differed significantly among stages (ANOVA: $F_{12,71} = 35.230$, $p < 0.001$) and between penaeids and sergestids ($F_{1,71} = 17.991$, $p < 0.001$), and there was also a significant interaction between the effects of stage and superfamily on these proportions ($F_{9,71} = 2.101$, $p = 0.041$) (Figure 2).

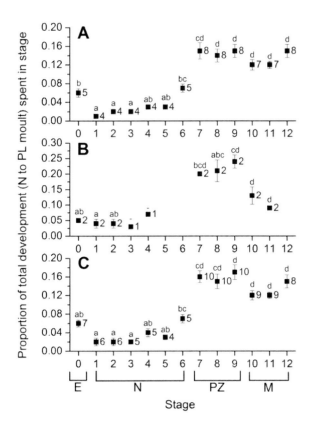

Figure 2. Mean ± SE proportions of total development (i.e., from start of the N phase to end of the M phase) taken up by different stages of (A) Penaeidae, (B) Segestidae, and (C) all Dendrobranchiata. Numbers shown beside each symbol are the number of species included in the calculation of the estimated proportion presented. Proportions significantly differed among stages for the Penaeidae (one-way ANOVA: $F_{12,63} =$ 34.300, p < 0.001), Segestidae ($F_{7,8} = 19.776$, p < 0.001, with the 3rd and 4th nauplius stages (n = 1 each) excluded), and all Dendrobranchiata ($F_{12,81} = 32.493$, p < 0.001). Different letters above symbols indicate stages with significantly different developmental proportions (Tukey's HSD test, p ≤ 0.05). E = embryo, N = nauplius, PZ = protozoea, and M = mysis. Stages labelled as 7-9 and 10-12 are stages PZ1-PZ3 and M1-M3, respectively.

In general, the first 5 nauplius stages were significantly shorter than all other stages, while the embryonic (E) phase preceding them and the subsequent 6th nauplius stage were similar in length to each other and longer than the first 5 nauplius stages (Figure 2A-C). The 3 protozoea stages were always similar in duration to each other, but longer than almost all other stages (Figure 2A-C). Among penaeids and for dendrobranchiates overall the proportion of development spent in the mysis stages did not significantly differ from that spent in the protozoea stages, although the first mysis stage did tend to be shorter than the preceding protozoea stages and the subsequent mysis stages became progressively longer (Figure 2A, C). Among the sergestids examined, the mysis stages were actually shorter than the protozoea stages, and the second mysis stage was shorter than the first, although both of these differences were not statistically significant (Figure 2B).

3.3. Actual Proportions of Pleocyematan Development Spent in Different Phases

Among the 8 species of Pleocyemata from different infraorders examined, there was considerable variation in the proportion of total development at comparable conditions (temperatures and salinities) spent in different identifiable phases (Table 2). When studies reported on the occurrence of eye pigment in embryos, it was possible to distinguish the durations of the combined E and N phases (extrusion to pigmentation) and the protozoea phase (pigmentation to hatching); this was not possible for two of the species examined [*Carcinus maenas* (Linnaeus, 1758) (Brachyura Latreille, 18028) and *Upogebia africana* (Ortmann, 1894) (Gebiidae de Saint Laurent, 1979)], so for these only the total duration of pre-mysis development (i.e., total embryonic period) could be considered (Table 2). The achelate lobster *P. argus* and the stenopodid shrimp *S. hispidus* spent considerably less time in their pre-mysis phases (6.7% and 15.6%, respectively) and much longer in the mysis phase (93.3% [assuming the achelatan phyllosoma larva is equivalent to a highly-modified mysis] and 84.4%, respectively) than did the dendrobranchiates or any other taxon examined (Table 2). The astacidean *H. americanus*, the caridean shrimp *Pandalus borealis* Krøyer, 1838 (Caridea Dana, 1852), and the gebiid

U. africana spent less time in the mysis phase and more time in the E, N, and PZ phases than did the dendrobranchiates, while all other taxa spent a greater proportion of development in the mysis stage and less time in the preceding phases than the dendrobranchiates (Table 2).

H. americanus spent a comparable proportion of development in the combined E and N phases (ca. 20%) to the dendrobranchiate average value of approximately 19% (Table 2); however, this species spent a remarkably greater proportion of time (up to 69%) in the PZ (i.e., embryo with pigmented eyes) phase than the dendrobranchiates and all other taxa (Table 2). The brachyuran *C. maenas* exhibited a uniquely plastic developmental pattern, wherein the proportion of development spent in different phases varied among temperatures: at temperatures ≤ 15°C, the pre-mysis phases took up the majority (up to ~63%) of development, whereas at higher temperatures the mysis stage encompassed a greater proportion (also up to ~63%) (Table 2). This variation suggests that dendrobranchiate-derived developmental proportions may not apply very well to estimating egg development times and their components of species in the Pleocyemata; nonetheless, this was still attempted and tested, as described below.

3.4. Estimation of Duration and Number of Embryonized Larval Stages in the Pleocyemata

Assuming that the timing of different embryonized phases within pleocyematan decapods could be estimated based on the data in Table 3, it is theoretically possible that the timing of stages within each phase could be predicted based on the proportions presented in Figure 2. As an example, this was attempted for the clawed lobster *H. americanus* (Astacidea), and the resulting predictions of the timing of different potential embryonized larval stages in this species are summarized within Table 4.

3.5. Predicted Pleocyematan Phase Durations and Comparisons to Observed Durations

Based on dendrobranchiate data, the duration of pre-mysis development, which is presumed to be equivalent to the combined duration of embryonized E, N, and PZ phases in the Pleocyemata, could potentially be estimated as being 1.65, 3.84, or 2.28 times the total duration of the mysis phase under a given condition (e.g., temperature) based on data for penaeids, sergestids, or all dendrobranchiates, respectively (Table 3); these results are shown in Figure 3.

Table 4. The cumulative time (in d) required for an American lobster, *Homarus americanus* (Pleocyemata: Astacidea), to reach different dendrobranchiate-equivalent development stages at different constant rearing temperatures

T	Phase	E[a]	N						PZ			M		
(°C)	Stage	0	1	2	3	4	5	6	1	2	3	1	2	3
5	Pen	56	67	78	101	134	168	235	392	538	683	818	952	1120
	Ser	56	90	134	168	246	246	246	448	650	896	1030	1120	1120
	Den	56	78	90	112	146	179	235	403	560	717	851	963	1120
10	Pen	14	16	19	25	33	41	57	96	131	167	199	232	273
	Ser	14	22	33	41	60	60	60	109	158	218	251	273	273
	Den	14	19	22	27	35	44	57	98	137	175	207	235	273
15	Pen	8	10	11	14	19	24	34	56	77	98	118	137	161
	Ser	8	13	19	24	35	35	35	64	93	129	148	161	161
	Den	8	11	13	16	21	26	34	58	81	103	122	138	161
20	Pen	6	7	8	10	13	17	24	39	54	68	82	95	112
	Ser	6	9	13	17	25	25	25	45	65	90	103	112	112
	Den	6	8	9	11	15	18	24	40	56	72	85	96	112
25	Pen	4	5	6	8	10	13	18	29	40	51	61	71	84
	Ser	4	7	10	13	18	18	18	34	49	67	77	84	84
	Den	4	6	7	8	11	13	18	30	42	54	64	72	84

Note: [a] E embryo, N = nauplius, PZ = protozoea, and M = mysis.
Estimated based on the proportions of development spent by dendrobranchiate species (Pen = Penaeidae, Ser = Sergestidae, and Den = all Dendrobranchiata) in each stage (see Figure 2). Observed durations were those reported by Perkins (1972) at each of the indicated temperatures. [E + N] = pre-pigmented embryo, PZ = pigmented embryo, M = larvae, and hatch should happen at the transition from PZ3 to M1

Likewise, the duration of pre-protozoeal development (from egg extrusion to pigmentation) and protozoeal duration (pigmentation to hatch) could be estimated as 0.487, 0.857, or 0.559 and 1.02, 3.02, or 1.35 of mysis duration, respectively (Table 2, 3). The results of these estimates are shown in Figure 4 and 5, and described for each species below.

For P. argus (Acheleta), the estimated total embryonic duration (Figure 3), duration of the pre-pigmented phases [E + N] (Figure 4), and length of the pigmented (PZ) phase (Figure 5) were markedly overestimated (by 80-656 d) using all dendrobranchiate groups' developmental ratios.

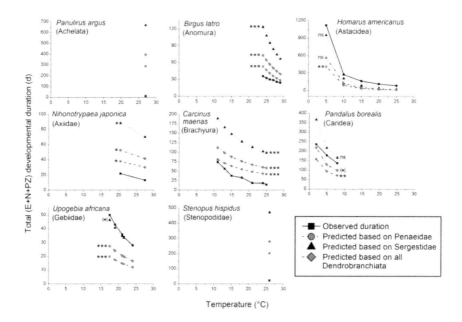

Figure 3. Observed (solid lines and black squares) and predicted (dashed lines and other symbols) total embryonic (presumed equivalent to [E + N + PZ]) durations (in d, y-axes) of species from different infraorders of Pleocyemata at different temperatures (°C, x-axes). Predictions were made based on observed larval (M) durations of the species presented here and the developmental proportions ([E + N + PZ]/M) in Table 3 for penaeids, sergestids, and all dendrobranchiates. For the sources of these data, see Table 2. It should be noted that the scales of the y-axes differ among species. Results of paired *t*-tests for each species comparing observed versus different types of predicted durations across temperatures are shown at the right or left end of the lines for predicted durations as follows: n.s.: not significant (p > 0.10); (*): marginally non-significant (0.05 < p ≤ 0.10); *: p ≤ 0.05; **: p ≤ 0.01; ***: p ≤ 0.001.

For *Birgus latro* (Linnaeus, 1767) (Anomura MacLeay, 1838), all dendrobranchiate-derived proportions significantly overestimated PZ (paired *t*-tests, p ≤ 0.05; Figure 3) and total embryonic durations at all temperatures (Figure 5). Durations of the [E + N] phases for *B. latro* were significantly underestimated using penaeid proportions and overestimated using sergestid proportions with marginal non-significance (paired *t*-tests, 0.05 < p ≤ 0.10), but those predicted based on overall dendrobranchiate data did not significantly differ from observed values (paired *t*-tests, p > 0.10; Figure 4).

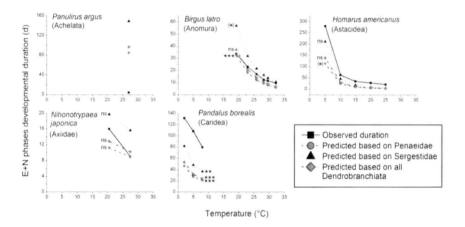

Figure 4. Observed (solid lines and black squares) and predicted (dashed lines and other symbols) durations of the pre-pigmented (presumed equivalent to [E + N]) embryonic phases (in d, y-axes) of species from different infraorders of Pleocyemata at different temperatures (°C, x-axes). Predictions were made based on observed larval (M) durations of the species presented here and the developmental proportions ([E + N]/M) in Table 3 for penaeids, sergestids, and all dendrobranchiates. For the sources of these data, see Table 2. It should be noted that the scales of the y-axes differ among species. Results of paired *t*-tests for each species comparing observed versus different types of predicted durations across temperatures are shown at the right or left end of the lines for predicted durations as follows: n.s.: not significant (p > 0.10); (*): marginally non-significant (0.05 < p ≤ 0.10); *: p ≤ 0.05; **: p ≤ 0.01; ***: p ≤ 0.001.

All predicted embryonic durations for *H. americanus* (Astacidea) were underestimated using dendrobranchiate proportions, especially at low temperatures, but the difference between observed and predicted durations was only significant for penaeid-derived predictions of total embryonic

(Figure 3) and PZ durations (Figure 5), and was marginally non-significant for the same predictions of [E + N] duration (Figure 4).

Total embryonic duration of *Nihonotrypaea japonica* Ortmann, 1891 (Axiidae de Saint Laurent, 1979) was significantly overestimated (Figure 3), and PZ duration was marginally non-significantly underestimated (Figure 5), using all dendrobranchiate groups' proportions; however, all predicted [E + N] durations of this species were not significantly different from observations (Figure 4).

Total embryonic duration of *C. maenas* (Brachyura) was significantly overestimated using penaeid, all dendrobranchiate, and especially sergestid developmental proportions (Figure 3).

Total embryonic duration of *P. borealis* (Caridea) was significantly underestimated by penaeid proportions and marginally non-significantly underestimated by overall proportions, while sergestid proportions produced overestimates that were not significantly different from observed values (Figure 3). All dendrobranchiate groups' proportions significantly underestimated [E + N] durations for *P. borealis* (Figure 4). PZ durations for *P. borealis* were significantly underestimated using penaeid proportions and significantly overestimated using sergestid proportions, but overall dendrobranchiate proportions produced predictions that did not significantly differ from observations (Figure 5).

Total embryonic durations of U. africana (Gebiidae) were significantly underestimated using both penaeid and overall dendrobranchiate proportions (Figure 3). Estimates based on sergestid proportions were very close to observed values, but were marginally non-significantly different from them (Figure 3).

Lastly, for S. hispidus (Stenopodidae) all dendrobranchiate-based proportions greatly overestimated total embryonic duration by 180-258 d (Figure 3).

The average ± SE absolute differences between predicted and observed total embryonic durations, averaged across pleocyematan infraorders and temperatures, were 74.22 ± 24.87 d (161.20 ± 74.59%) for penaeid proportions, 98.00 ± 24.27 d (399.28 ± 179.98%) for sergestid data, and 77.23 ± 21.88 d (227.84 ± 104.82%) for all dendrobranchiate data. For the pre-pigmented [E + N] phase the differences were 35.40 ± 11.06 d (148.94 ± 102.34%), 36.59 ± 8.87 d (228.54 ± 186.63%), and 32.90 ± 10.12 d (160.10 ± 119.00%) for penaeid, sergestid, and dendrobranchiate data, respectively, and for the pigmented (PZ) phase these differences were 74.76 ± 34.51 d (215.68 ± 121.05%), 113.84 ± 27.61 d (704.92 ± 370.35%), and 75.49 ± 30.35 d (292.34 ± 162.36%).

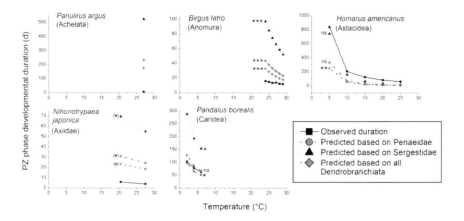

Figure 5. Observed (solid lines and black squares) and predicted (dashed lines and other symbols) durations of the pigmented (presumed equivalent to PZ) embryonic phases (in d, y-axes) of species from different infraorders of Pleocyemata at different temperatures (°C, x-axes). Predictions were made based on observed larval (M) durations of the species presented here and the developmental proportions (PZ/M) in Table 3 for penaeids, sergestids, and all dendrobranchiates. For the sources of these data, see Table 2. It should be noted that the scales of the y-axes differ among species. Results of paired *t*-tests for each species comparing observed versus different types of predicted durations across temperatures are shown at the right or left end of the lines for predicted durations as follows: n.s.: not significant (p > 0.10); (*): marginally non-significant (0.05 < p ≤ 0.10); *: p ≤ 0.05; **: p ≤ 0.01; ***: p ≤ 0.001.

Across all taxa of Pleocyemata, the durations of total embryonic development predicted using penaeid (paired *t*-test: t_{29} = 0.984, p = 0.333) and all dendrobranchiate (t_{29} = -0.002, p = 0.998) proportions did not significantly differ from observed durations, but predictions based on sergestid proportions significantly overestimated observed durations (t_{29} = -2.481, p = 0.019). The combined duration of the [E + N] phases did not significantly differ between observed and predicted values (Penaeidae: t_{16} =

2.061, p = 0.056; Sergestidae: t_{16} = 0.455, p = 0.655; All: t_{16} = 1.803, p = 0.090), although penaeid and overall dendrobranchiate predictions were marginally non-significantly lower than observations. Predicted PZ durations did not significantly differ from observed durations overall (Penaeidae: t_{16} = 1.155, p = 0.265; Sergestidae: t_{16} = -1.915, p = 0.074; All: t_{16} = 0.735, p = 0.473), although sergestid-based proportions did overestimate observed durations with marginal non-significance.

Overall, then, developmental proportions based on data for any dendrobranchiate groups' free-living early larval phases did not perform well at predicting the duration of embryonic development (in total or either (pre-pigmented = [E + N], pigmented = PZ) of its major sub-stages) for any of the species of Pleocyemata examined.

4. Discussion

4.1. Can Dendrobranchiate Development Be Used to Predict Embryonic Durations of Pleocyemata?

Based on the results of analyses performed in this chapter, it can be concluded that the proportions of development spent by penaeid and sergestid dendrobranchiates in different free-living larval phases and stages cannot be used to accurately predict the embryonic durations of species of Pleocyemata. Indeed, embryonic development times of Pleocyemata predicted based on dendrobranchiate development differed considerably from observed durations reported in the literature, especially for certain species (e.g., *H. americanus*, *P. argus*, and *S. hispidus*). However, a comparison between the timing of different stages and phases in the embryonic development of *H. americanus* estimated in this chapter (Table 4) and that of putative phases presented by Talbot and Helluy (1995) illustrates that this discrepancy may not be so extreme overall. Talbot and Helluy (1995) identified the naupliar phase as beginning (with the embryo starting to "twitch") just before 10% of total embryonic development had been completed and lasting until embryonic development was ~20%

completed, after which the remaining ~80% of embryonic development was composed of the eyed "prelarva" phase. In this chapter, the nauplius phase was predicted to begin when 6-8% of pre-mysis development was completed and last until 28-34% of development was completed, and then the protozoea phase took up the remaining 66-72% of the embryonic period (Figure 2 and Table 4, values calculated by dividing stage proportions by the cumulative proportion of time to the end of PZ3). Therefore, predictions for *H. americanus* were actually not so far from realistic values, although the length of the nauplius phase was overestimated and that of the protozoea phase underestimated by ~10% herein relative to Talbot and Helluy (1995). That said, at very low temperatures this discrepancy led to large prediction errors, which shows that this method was still not optimal.

The above indicates that alternative methods should likely be pursued to improve hatch timing prediction among the "higher" decapod crustaceans. However, the present analyses were only a first attempt at making such predictive inferences, and were limited by a number of methodological (e.g., data quality and availability) and biological (e.g., unquantified sources of variability among taxa) limitations. Through this attempt, a number of lessons were learned that can be applied in future studies, and set the stage for potential future analyses of differences in developmental proportions among and within crustacean taxa. Additionally, some interesting aspects of immature development in the species studied were revealed during these analyses that may bear further study. These lessons are the focus of most of the following Discussion.

4.2. General Methodological Limitations to the Analyses Performed

The number of species of dendrobranchiates for which data could be obtained was relatively small (13), and the data obtained were limited to those from members of only two (Penaeidae and Sergestidae) of the seven families in the Dendrobranchiata. Development in other taxa could have differed from that of the two families examined. For example, the

Solenoceridae and Benthesicymidae are primarily deep-water taxa of Penaeoidea, which have 5-6 nauplius, 3 protozoea, and 2 or 4 mysis stages, respectively (Martin et al. 2014a). It is also reasonable to expect that there might be more intra-family variation in development (i.e., among genera or species of Penaeidae) than was captured in the data collected. Members of the sergestoid family Luciferidae are unique among the Dendrobranchiata in that they brood their embryos for a short period (Martin et al. 2014a), which may imply evolutionary similarity to the Pleocyemata, who brood their embryos until they hatch into larvae; it would be interesting to see how the developmental proportions of luciferid larvae compare with those of the Pleocyemata. Ultimately, a more elaborate analysis including more species and families of Dendrobranchiata may produce better estimates of the proportions of development spent by these prawns in different developmental phases and stages, which might be able to make more reasonable estimates of pleocyematan embryonic durations (but see below). It is not clear whether this is possible at present, because the larval development of members of other dendrobranchiate groups besides the Penaeidae and Sergestidae may not have been studied as extensively as that in these groups due to them occupying relatively inaccessible habitats (e.g., the deep sea) in nature (Martin et al. 2014a). Development time data for penaeid and sergestid species other than those examined herein may also not be available, and/or good quality data permitting strong statistical tests may not yet be available for such species. If additional data for such taxa could be obtained, however, then a formal review and/or phylogenetic analysis of developmental proportions among all major dendrobranchiate taxa (i.e., heterochrony analysis) might be conducted, which could reveal: (1) how much variation in developmental timing of different phases and stages occurs within this decapod suborder; and (2) which, if any, developmental pattern is the most ancestral for the group, and thus might be closer to the expected ancestral developmental pattern of the Pleocyemata.

It is also worth noting that only a small number of species of Pleocyemata were used in tests of the predictive ability of dendrobranchiate proportions, with one species used each from 8 out of 11 of the infraorders within this suborder. The species used to represent each infraorder were selected somewhat arbitrarily, and thus it is possible (and in the case of *H. americanus* quite probable) that they represented outliers for the developmental programs in their respective taxa. A more extensive analysis could be conducted using multiple species from each infraorder, ideally with multiple representative species from different taxa within these groups included to better capture the developmental variability within as well as among taxa of Pleocyemata. While it is not currently possible to increase the number of infraorders for use in such an analysis due to the limited or nonexistent information on the embryonic development and larvae of the Glypheidea, Polychelida, and Procaridiidea (Martin 2014b), future studies may provide data that will make this possible as well.

An important limitation to all analyses performed, especially the assessment of developmental proportions of Dendrobranchiata, was the low quality of the data obtained and resultant low sample sizes and limited ability to perform statistical analyses. For most dendrobranchiate taxa and also some Pleocyemata (e.g., *S. hispidus*), only a single value for the duration of each and/or all developmental phases and stages could be obtained from the literature. It is clearly impossible to know under such circumstances whether such a value is representative of the development of a given species, and this does not allow one to have any idea of the amount of intraspecific variation in development times. This also meant that a formal statistical analysis of the interspecific variation in the proportion of development spent in each phase or stage among the Dendrobranchiata (and taxa within this group) could not be performed.

If all studies obtained had reported mean development times along with error measures (e.g., SE) and sample sizes, then it would have been possible to extract multiple replicate durations for each species, stage, and experimental condition, giving a representation of the variation in developmental proportions within each species. Then, an analysis (e.g., ANOVA) of whether developmental proportions for each phase or stage varied significantly among species (and/or genera, families, superfamilies, etc.) could have been performed. Optimally, this step would have preceded the use of dendrobranchiate proportions to predict embryonic development in Pleocyemata, as significant variability within the Dendrobranchiata would indicate the likely inability of their developmental proportions to predict development in Pleocyemata. Likewise, if replicate data for all taxa of Pleocyemata used for the comparison of predicted versus observed development times could have been obtained, this would have allowed for more powerful tests of the predictive ability of dendrobranchiate proportions, as well as perhaps differences in development among Pleocyemata. This was not possible in the analyses conducted in this chapter. However, perhaps a more extensive literature search could be conducted in the future, for example using different article databases and search terms and delving further into the non-English and secondary literature, which would allow for more and perhaps better data to be acquired. Such a search and detailed follow-up analyses should be conducted before completely rejecting the possibility that the methods examined herein could be useful for predicting embryonic development times and stages in the pleocyematan decapods.

Analyses performed herein also did not consider the decapodid ("postlarval") phase of development in either dendrobranchiates or the Pleocyemata, due to the well-known variability in the length and number of stages in this phase in some taxa (Felder et al. 1985; Oliphant et al. 2013; Quinn 2016; Williamson 1982) and the uncertainty of its stage numbers or ending point in Dendrobranchiata.

However, this settling phase is a demographically important component of immature development in nearly all decapods (Felder et al. 1985), and may be one of the longest parts of the larval phase in terms of duration (e.g., in *H. americanus* the duration of the decapodid stage is potentially equal to or slightly more than that of all three mysis stages combined; MacKenzie 1988). Therefore, in a future analysis it may be worthwhile to also examine the proportions of different species' development spent in this phase, as well as to maybe attempt to predict pleocyematan decapodid duration based on dendrobranchiate developmental proportions for this phase.

4.3. Potential Limitations due to Biological Variability among Taxa

Three major overarching assumptions made in the analyses and inferences performed in this chapter were that: (1) the free-living nauplius and protozoea larvae of dendrobranchiates are homologous to the "egg-nauplius" and pigmented embryonic phases, respectively, of Pleocyemata; (2) significant evolutionary shifts in the timing of events during development, called heterochrony, has not occurred among species between or within either decapod suborder; and (3) the proportion of total immature development spent in each stage does not vary within the same species (i.e., development is equiproportional). The validity of each of these assumptions and their potential impacts on the conclusions herein are discussed in the following paragraphs.

The (free-living) nauplius is assumed to be an ancestral shared feature of all crustaceans that has been lost in some groups (e.g., Pleocyemata), with the nauplius representing either the larval form of the ancestral crustacean or even being the ancestral crustacean itself (Dahms 2000; Ferrari et al. 2011; Martin et al. 2014c). Some authors have debated this, however, claiming that all or some crustacean nauplii were secondarily and separately evolved (Scholtz 2000) or acquired by different groups (Williamson 1992, 2003, 2006), although this latter view has not been confirmed or accepted by most scholars. There has been much more debate in the literature over whether the nauplii of dendrobranchiate (and euphausiid) species are truly homologous to the nauplii seen in other, non-malacostracan crustaceans (Scholz 2000), as well as whether the "egg-nauplius" of the Pleocyemata truly represents an evolutionary equivalent to a free-living nauplius (Jirkowski et al. 2013, 2015; Williamson 2006). Nauplii of dendrobranchiates and euphausiids are different in many respects from those of other crustaceans, as for example they are lecithotrophic and do not feed whereas in most other taxa the naupliar stages after the first one usually do feed (Akther et al. 2015; Martin and Gómez-Gutiérrez 2014; Martin et al. 2014b; Williamson 1982). They were thus previously thought to have been secondarily evolved in these groups after they were lost in their ancestors

(Scholtz 2000), meaning that they are not homologous to other crustaceans' nauplii, or perhaps even to the "egg-nauplii" of related species. Williamson (2006) suggested that the ancestors of crustaceans without nauplii never had them, which would mean that the dendrobranchiates and euphausiids acquired their nauplii secondarily (e.g., via hybridization-mediated "larval transfer": Williamson 1992, 2003) and that the "egg-nauplius" of the Pleocyemata is not equivalent to a nauplius, and thus dendrobranchiate nauplii and pleocyematan "egg-nauplii" are not homologous. While all of these debates remain largely unsettled, recent studies that focused on morphology (Akther et al. 2015; Jirkowsky et al. 2013) and gene expression (Jirkowsky et al. 2015) during development have provided strong evidence that the dendrobranchiate and euphausiid nauplius stages are indeed produced by the same developmental processes that are active in the "egg-nauplius" of some taxa, and in the free-living nauplii of non-malacostracan taxa; differences between these groups' nauplii and those of others appear to be mainly due to adaptations to their non-feeding lifestyle (Akther et al. 2015). Therefore, based on the above discussion there is reasonable evidence in support of equating the free-living dendrobranchiate nauplius stages and the "egg-nauplius" phase of the Pleocyemata.

The homology between dendrobranchiate protozoeae and the phase of embryonic development in Pleocyemata that bears pigmented lateral eyes has not been studied as much as that between nauplii and "egg-nauplii". The acquisition of the lateral compound eyes is a likely indicator of the transition between a nauplius-like phase and a protozoea-like one (Martin et al. 2014a; Williamson 1982). Some pleocyematan species are also occasionally observed to hatch as a pre-mysis stage without the ability to swim using its thoracic appendages, referred to as a "prezoa" or "prelarva" in H. americanus (Talbot and Helluy 1995) and a "naupliosoma" in achelate lobsters (Phillips and Sastry 1980), which then quickly moults into the normal first larval (mysis) stage; this hatchling stage could be considered to represent the last protozoea stage, although this does not appear to have been investigated in any detailed, formal analyses. Although they did not discuss the protozoeae specifically in much detail, Jirkowski et al. (2015) did observe that developmental events that occur during this phase in

Dendrobranchiata also occur after the "egg-nauplius" phase and before hatching in Pleocyemata, which provides some support for concluding homology between at least the early part of the pigmented embryo phase in Pleocyemata and dendrobranchiate protozoeae; however, this should be confirmed in future studies on this life history phase.

Heterochrony is known to have played a significant role in the evolution of many different organisms, including for example vertebrates (McNamara 2012) and species of lobsters (Rötzer and Haug 2015), as genetic or epigenetic changes in the timing of developmental events can lead to new life histories that experience different evolutionary pressures (McNamara 2012). This process is already well established to have occurred among taxa of decapods and other malacostracan crustaceans (Gore 1985; Jirkowsky et al. 2015; Rötzer and Haug 2015), where it has led to the loss or reacquisition, abbreviation or prolonging, and morphological differentiation of equivalent larval stages among related species (Gore 1985; Williamson 1982). Heterochrony can lead to one taxon spending a relatively longer proportion of its developmental period in one phase or stage and thus less time in others, or vice versa, in comparison to another related taxon. Therefore, the degree of heterochrony within and especially between the Dendrobranchiata and Pleocyemata is an important limitation to whether and how effectively dendrobranchiate developmental proportions can be used to predict pleocyematan embryonic development times.

While variations in developmental proportions for each phase and stage were fairly small among the Dendrobranchiata (although those for the protozoea and mysis phase differed between penaeids and sergestids), differences among the species of Pleocyemata examined were quite large. More importantly, the differences between the proportions of development spent in all equivalent phases differed between the dendrobranchiate and pleocyematan species examined, in some cases to extreme degrees. This indicates that there have probably been substantial heterochronic shifts in the timing of developmental phases between the Pleocyemata and their last shared common ancestor with the Dendrobranchiata, and there appears to have been considerable heterochrony in the evolutionary transitions among infraorders of Pleocyemata as well. Given this, dendrobranchiate development proportions would not be expected to predict pleocyematan embryonic development times very well, if at all. However, as mentioned earlier these findings are preliminary, and should be confirmed with larger sample sizes, more thorough sampling of different decapod taxa, and using formal analyses informed by phylogeny.

Using the observed duration of one developmental stage with the relative ratios or proportions of development spent in each stage to predict the duration of another stage (as done in this chapter) draws from previous work performed extensively on copepod crustaceans (Corkett 1984; Corkett and McLaren 1970; Hart 1990, 1998), as well as on the lobster *H. americanus* (MacKenzie 1988). A certain school of thought among researchers working on these crustaceans proposes that for some (or perhaps all) species the proportion of total immature development spent in each larval phase or stage is a species-specific and constant value (Corkett 1984). This type of development is referred to as equiproportional development, or EPD (Hart 1990, 1998). If a species has EPD, then the duration of one developmental stage under any (non-extreme) rearing conditions, for example at different temperatures, can be estimated based on prior knowledge of the species' developmental ratios or proportions (Corkett 1984; Corkett and McLaren 1970). However, in a recent review (Quinn 2018) that tested temperature-dependent development data from a number of arthropod species for the occurrence of EPD, including copepod and decapod crustaceans, it was found that almost no species truly has EPD, even those species previously reported as having such. Although EPD was able to provide reasonable estimates of stage durations for most species and temperatures tested, a far better-supported model of development that had better predictive ability was found to be one in which developmental proportions varied with temperature: i.e., the proportion of total development of a species spent in each stage varies among temperatures, but is a constant *function* (linear, or more often quadratic) of temperature (Quinn 2018). Indeed, another review (Quinn 2016) revealed the fact that many decapod species can alter the duration, number, and types of stages through which they develop depending on the conditions they experience (e.g., Oliphant et al. 2013), and thus do not have EPD. Therefore, this developmental plasticity may possibly represent a general phenomenon among crustaceans or arthropods as a whole (Quinn 2016, 2018).

Predictions and analyses made in this chapter implicitly assumed that all taxa had EPD, largely as a simplifying assumption, but given the recent findings of Quinn (2018) this seems unlikely to be realistic. Indeed, the fact that analyses in this chapter found that the crab *C. maenas* spent more time in its embryonic than larval phase at temperatures ≤ 15°C, equal proportions at ~15°C, and more time in the larval than embryonic phase at temperatures > 15°C, is a clear sign that this species, at least, does not have EPD. Because of limited sample sizes and the unavailability of both embryonic and larval durations at multiple temperatures for some species of both Dendrobranchiata and Pleocyemata, it was not possible to account for the temperature-dependence of developmental proportions of many species herein, so phase and stage proportions were averaged across different rearing temperatures for each species. Therefore, some of the errors in predictions may have been related to temperature-mediated variations in actual developmental proportions. Future studies attempting analyses like those done in this chapter should try to account for variability in the developmental proportions of each species due to rearing conditions like temperature, although it should be noted that such developmental variability might make it difficult or impossible to compare developmental proportions among species.

4.4. Developmental Observations of Biological Interest

The developmental proportions calculated herein for each phase and stage were relatively consistent among most of the dendrobranchiate species examined, but varied greatly both between these and the pleocyematan species examined and among the Pleocyemata. This may indicate that developmental patterns among the Dendrobranchiata are fairly uniform, with little heterochrony having occurred within this taxon in terms of phases even among species in which one or more nauplius or mysis stage appears to have been lost (e.g., the sergestids). On the other hand, much heterochrony appears to have occurred between them and the Pleocyemata, and among pleocyematan infraorders. This conclusion should be taken with caution, however, as the vast majority of dendrobranchiates examined came from two closely related genera within one family: 5 species each of *Metapenaeus* and *Penaeus* of the family Penaeidae. At the very least, this indicates conservation of developmental programs within this family and/or these genera. A more extensive analysis of dendrobranchiate development, including members from all 7 families, is needed to confirm the apparent uniformity within this group. If confirmed, however, it may point to some interesting evolutionary constraints or a lack of diversity or genetic variation within the group, which has evolutionary implications. An analysis including the Euphausiacea, the only other malacostracan group other than the Dendrobranchiata that develops through four free-living larvae phases including a nauplius phase (Akther et al. 2015; Martin and Gómez-Gutiérrez 2014; Ross 1981; Williamson 1982) may also be informative. The evolutionary factors responsible for the "embryonization" of larval stages in the transition between the Dendrobranchiata and Pleocyemata also remain unclear, but should be investigated.

Among the Dendrobranchiata, the proportion of development spent in successive stages within the nauplius and mysis phases tended to increase, and there was also a dramatic increase in the proportion of time spent in the protozoea and mysis phases and stages relative to those spent in the naupliar ones. This increase in relative duration among successive stages agrees with patterns previously observed for pleocyematan decapods (e.g., *H. americanus*: MacKenzie 1988), as well as many different taxa of crustaceans and some other arthropods (Anger 2001; Quinn 2018). This may be related to the simple fact that after the moult transitioning from one stage into the next one, a larger animal is created, which will require more time to grow and modify its morphology before it can moult again (Anger 2001; Gore 1985). The fact that the last stage in a phase also tends to be the longest (with the exception of the protozoea stages, which seem to be nearly isochronal to each other) also agrees with previous observations in other taxa (Anger 2001; MacKenzie 1988; Quinn 2018). This is likely due to the fact that the moult following this final stage will entail a metamorphosis, in which massive changes to morphology occur (Felder et al. 1985; Gore 1985; Williamson 1982). It is reasonable to expect that it should take additional time for the animal to prepare a larger and more complex body, with new appendages, segments, musculature, etc., before it is ready to complete a metamorphic moult. The last (sixth) nauplius stage of penaeids stood out as being particularly long relative to preceding stages, perhaps due to the impending transformation from a nauplius to a protozoea. The last mysis stage also precedes a metamorphosis, that to the decapodid or "postlarva" phase in which larval morphology is discarded and the juvenile/adult form appears (Felder et al. 1985; Phillips and Sastry 1980). The change in morphology from protozoea to mysis is comparatively small, which may explain why the duration of the last protozoea stage is not apparently longer than that of other stages in this phase. Increasing relative durations thus may simply have resulted from constraints to growth, moulting, and development rates, although these also have ecological implications in variable natural temperature regimes. Most decapod species release their larvae early in the spring or summer, meaning that the earliest developmental stages, which are relatively shorter at a particular temperature, experience lower temperatures than they will later in the season as later-stage larvae, which are relatively longer in duration at a given temperature; thus, the longest stages experience the warmest conditions, which reduces their duration. This process may mean that in nature, under fluctuating thermal regimes, different developmental stages are nearly isochronal (i.e., take the same length in time), which may allow decapod larvae to optimize their development times.

The large variation in developmental proportions among species of Pleocyemata observed may be due to extensive phylogenetic divergence among infraorders and/or evolutionary specializations within these different taxa to deal with different environmental pressures. *H. americanus* is known to have one of the longest incubation periods of any lobster or decapod (MacDiarmid and Sainte-Marie 2006), and thus its larval period is proportionally one of the shortest relative to its embryonic period. Perhaps this long incubation period functions to improve the size, provisioning, and condition of fewer larvae upon hatching, and thus improves larval survival within the highly competitive and predation-heavy coastal plankton they inhabit (MacDiarmid and Sainte-Marie 2006; Phillips and Sastry 1980). The fact that this species spent the greatest proportion of its developmental period within the pigmented embryo (PZ?) phase was an interesting finding that stood out, although the significance of this finding is not clear at present; this will bear further investigation. Achelate lobsters, including *P. argus*, have some of the longest larval durations of any crustacean (MacDiarmid and Sainte-Marie 2006; Phillips and Sastry 1980) and relatively short embryonic periods. Many species inhabit reefs or isolated seamounts as adults, but spend their larval life entrained in large-scale gyres of the open ocean (Phillips and Sastry 1980). They release many small phyllosoma larvae that are uniquely adapted to life in the water column of the open ocean, and undergo considerable growth over their extended dispersal period. The short embryonic duration in these species may be related to their long larval life, and perhaps biotic pressures from the adult habitats. The priority in these species may be to release the larvae into the water column as soon as possible, to get them away from the crowded and predation-intensive adult habitats into a comparatively safe open ocean environment (MacDiarmid and Sainte-Marie 2006; Phillips and Sastry 1980), where the larvae then take ample time to feed, grow, and return to their natal habitats via the rotations of large oceanic gyres. Thus, the proportions of development spent in the embryonic and larval phases of achelate lobster have become extremely short and long, respectively. Interestingly, the ornamental cleaner shrimp *S. hispidus* (Stenopodidae) was also found to have a similarly reduced embryonic and extended larval period, which could suggest that it has been subjected to similar selective pressures as those impacting the achelate lobsters. The temperature-dependent developmental ratios of the brachyuran crab *C. maenas* were also an interesting finding, with implications to this species' role as an invasive species around the world (Nagaraj 1993), as being able to adjust its developmental pattern in the face of environmental heterogeneity can give a species an advantage over others under certain circumstances (Quinn 2016, 2018, and references therein). Differences in developmental proportions among the other species of Pleocyemata examined may similarly point to aspects of their habitats and ecological and evolutionary pressures on them, which should be investigated further.

4.5. Implications of Results to Identifying Embryonic Stages and Predicting Hatch Timing

The results of the coarse analyses presented in this chapter do not indicate that inferences based on dendrobranchiate development can be applied to predict the embryonic development times of "higher" decapods. In the preceding sections, ways in which these techniques might be improved and potentially lead to more useful inferences of pleocyematan development, or at least lead to other biological findings of interest, were discussed. Even if these approaches were improved and performed well at predicting literature data, an essential next step would be to validate them through the examination of real pleocyematan embryos. One approach to do so would involve dissection of embryos, either at regular intervals (daily or weekly) and/or at preselected time points corresponding to those when, based on dendrobranchiate-based inferences for the species and temperature under study, the embryo was expected to transition to a new "embryonized" larval stage. Some previous studies (e.g., Alwes and Scholtz 2006; García-Guerrero and Hendrickx 2004, 2006a, b; Müller et al. 2004, 2007; Talbot and Helluy 1995) have taken the former approach, and used it to identify potential embryonic stages. However, such efforts have produced somewhat vague definitions of developmental "stages" that gradually grade into one another, rather than undergoing the stepwise and moult-mediated transitions typically associated with developmental transitions in crustaceans (Forster et al. 2011; Quinn 2018; Williamson 1982). Based on observations made on the setae of dissected embryos, some studies have proposed that they have observed one or more embryonic moults, including naupliar moults (Talbot and Helluy 1995), but these observations remain largely unconfirmed.

Optimally, a description of embryonic stages should include detailed observations on the morphology (and/or perhaps the gene expression; Jirkowski et al. 2015) of each stage, complete with observations of the embryonic moults (if any) marking the transition between stages. Recent studies aiming to improve predictions of hatch timing for species of Pleocyemata have focused on refining techniques involving indirect indices of development, such as eye size (e.g., García-Guerrero M, Hendrickx 2004, 2006a, b; Haarr 2018; Miller et al. 2016; Ouellet and Chabot 2005), while the potential of basic, direct studies of the embryos themselves, such as via dissections, which may also lead to worthwhile insights, have been largely neglected. To truly improve our understanding of the embryonic development of "higher" decapods, however, this area of study needs to be revisited and refined. Electron microscopy could potentially be applied to greatly improve the detail with which the morphological features of different stages are described. New techniques may also be developed to allow for the observation of individual living embryos as they progress through development (and perhaps moults). The use of micro-video cameras, partial removal of the egg membrane to permit a view through a "window" into the inside of the egg, and/or development of some means to remove the egg membrane while keeping the embryo alive (and hopefully developing normally) are techniques that might someday be used to directly observed development of pleocyematan embryos in more detail than has been possible before. In the meantime, other avenues for improving predictions of embryonic development times and hatch timing should be developed, and are being developed for some species, including by assessing development rates of individually reared embryos (Giovagnoli et al. 2014; Hartnoll and Paul 1982; Sha Bo, University of New Brunswick, Saint John, NB, Canada, M.Sc. thesis in progress) and better assessing the sources of variability in development rates of embryos, including variability among embryos within a brood, among broods, and among geographic locations (i.e., putative locally-adapted subpopulations) (Haarr 2018).

Predicting hatch timing is an important source of data needed for models of ecological interactions, phenology, and larval dispersal and recruitment, as the timing of larval release can greatly impact the survival, condition, and destination of dispersing larvae (Haarr 2018; Miller et al. 2016; Pineda and Reyns 2018; Reisser et al. 2013; Shanks 2009) and the recruitment to adult populations that results from larval inputs (O'Connor et al. 2007; Phillips and Sastry 1980; Pineda and Reyns 2018). Methods used traditionally and currently to make such predictions, for example in the study of species supporting fisheries (Miller et al. 2016; Perkins 1972; Shumway et al. 1985), implicitly assume that embryonic development rate at a given temperature is constant across all embryonic stages. However, there is some evidence that this is not correct and that development rates change in a stepwise manner over the course of development (Lydia White, University of New Brunswick, Saint John, NB, Canada, pers. comm.), which may be an indicator of (1) the existence of different embryonic stages than those currently known and (2) variation in temperature-dependent development rates among such stages. This is important to predicting development rates and hatch timing, because different stages also tend to encompass different proportions of the developmental period, with some taking up much more time at a given temperature than others. Because developmental proportions and development rates vary among stages and temperatures, and because temperatures experienced by embryos in nature can vary considerably (MacDiarmid and Sainte-Marie 2006; Perkins 1972), predicting hatch timing is very complicated and is beset by many potential sources of error. Compared with predictions made assuming no variations in development among stages, the combination of a relatively long stage experiencing particularly low temperatures, for example, will greatly delay developmental progression (and subsequently hatching), while a relatively short stage experiencing high temperatures will greatly accelerate development. It is therefore essential that the different stages in the embryonic development of species of Pleocyemata be identified and their characteristics, particularly differing development times or rates, quantified.

ACKNOWLEDGMENTS

I thank the University of New Brunswick (UNB), Saint John Campus for providing access to literature and software that made this review possible, and Megumi Quinn for support during the preparation of the manuscript. I also thank Nadya Gotsiridze-Columbus and Nova Science Publishers, Inc., for inviting me to submit this chapter to the present collection.

REFERENCES

Akther H, Agersted MD, Olesen J (2015) Naupliar and metanaupliar development of *Thysanoessa raschii* (Malacostraca, Euphaususiacea) from Godthåbsfjord, Greenland, with a reinstatement of the ancestral status of the free-living nauplius in malacostracan evolution. *PLoS ONE* 10(12):e0141955. DOI:10.1371/journal.pone.0141955.

Alwes F, Scholtz G (2006) Stages and other aspects of the embryology of the parthenogenetic Marmokrebs (Decapoda, Reptantia, Astacida). *Dev Genes Evol* 216:169-184.

Anger K (2001) *Crustacean Issues 12: The Biology of Decapod Crustacean Larvae*. Rotterdam, The Netherlands: A.A. Balkema.

Bradbury IR, Snelgrove PVR, Fraser S (2000) Transport and development of eggs and larvae of Atlantic cod, *Gadus morhua*, in relation to spawning time and site in coastal Newfoundland. *Can J Fish Aquat Sci* 57:1761-1772.

Bressan CM, Müller YMR (1997) Characterization of embryonized nauplius development *Macrobranchium acanthurus* (Crustacea, Decapoda). *Braz J Morphol Sci* 14(2):243-246.

Brillon S, Lambert Y, Dodson J (2005) Egg survival, embryonic development, and larval characteristics of northern shrimp (*Pandalus borealis*) females subject to different temperature and feeding conditions. *Mar Biol* 147:895-911.

Churchill JH, Runge J, Chen C (2011) Processes controlling retention of spring-spawned Atlantic cod (*Gadus morhua*) in the western Gulf of Maine and their relationship to an index of recruitment success. *Fish Oceanogr* 20(1):32-46.

Corkett CJ (1984) Observations on development in copepods. In: *Crustaceana Supplement No. 7, Studies on Copepoda II (Proceedings of the First International Conference on Copepoda, Amsterdam, the Netherlands, 24-28 August 1981)*, pp. 150-153.

Corkett CJ, McLaren IA (1970) Relationships between development rate of eggs and older stages of copepods. *J Mar Biol Assoc UK* 50:161-168.

Crisp JA, Tweedley JR, D'Souza FML, Partridge GJ, Moheimani NR (2016) Larval development of the western school prawn *Metapenaeus dalli* Racek, 1957 (Crustacea: Decapoda: Penaeidae) reared in the laboratory. *J Nat Hist* 50(27-28):1699-1724.

Dahms H-U (2000) Phylogenetic implications of the Crustacean nauplius. *Hydrobiologia* 417:91-99.

Dall W, Hill BJ, Rothlisberg PC, Sharples DJ (1990) The biology of the Penaeidae. *Adv Mar Biol* 27:1-489.

Ewald JJ (1965) The laboratory rearing of pink shrimp, *Penaeus duorarum* Burkenroad. *Bull Mar Sci* 15(2):436-449.

Felder DL, Martin JW, Goy JW (1985) Patterns in early postlarval development of decapods. In: *Crustacean Issues 2: Larval Growth*. Wenner AM (ed.). Rotterdam, The Netherlands: A.A. Balkema, pp. 163-225.

Ferrari FD, Fornshell J, Vagelli AA, Ivanenko VN, Dahms H-U (2011) Early post-embryonic development of marine chelicerates and crustaceans with a nauplius. *Crustaceana* 84(7):869-893.

Fletcher DJ, Kötter I, Wunsch M, Yasir I (1995) Preliminary observations on the reproductive biology of ornamental cleaner prawns *Stenopus hispidus, Lysmata amboinensis* and *Lysmata debilus. Int Zoo Yb* 34:73-77.

Forster J, Hirst AG, Woodward G (2011) Growth and development rates have different thermal responses. *Am Nat* 178:668-678.

García-Guerrero M, Hendrickx ME (2004) Embryology of decapod crustaceans I. Embryonic development of the mangrove crabs *Goniopsis pulchira* and *Aratus pisonii* (Decapoda: Brachyura). *J Crustac Biol* 24(4):666-672.

García-Guerrero M, Hendrickx ME (2006a) Embryology of decapod crustaceans, II. Gross embryonic development of *Petrolisthes robsonae* Glassell, 1945 and *Petrolisthes armatus* (Gibbes, 1850) (Decapoda, Anomura, Porcellanidae). *Crustaceana* 78(9):1089-1097.

García-Guerrero M, Hendrickx ME (2006b) Embryology of decapod crustaceans III: Embryonic development of *Eurypanopeus canalensis* Abele & Kim, 1989, and *Panopeus chilensis* H. Milne Edwards & Lucas, 1844 (Decapoda, Brachyura, Panopeidae). *Belg J Zool* 136(2):249-253.

Gendron L, Ouellet P (2009) Egg development trajectories of early and late-spawner lobsters (*Homarus americanus*) in the Magdalen Islands, Québec. *J Crustac Biol* 29:356–363.

Giovagnoli A, Ituarte RB, Spivak ED (2014) Effects of removal from the mother and salinity on embryonic development of *Palaemonetes argentinus* (Decapoda: Caridea: Palaemonidae). *J Crustac Biol* 34(2):174-181.

Goldstein JS, Matsuda H, Takenouchi T, Butler MJ, IV (2008) The complete development of larval Caribbean spiny lobster *Panulirus argus* (Latreille, 1804) in culture. *J Crustac Biol* 28(02):306-327.

Gore RH (1985) Moulting and growth in decapod larvae. In: *Crustacean Issues 2: Larval Growth*. Wenner AM (ed.). Rotterdam, The Netherlands: A.A. Balkema, pp. 1-65.

Gregati RA, Fransozo V, López-Greco LS, Negreiros-Fransozo ML (2010) Reproductive cycle and ovarian development of the marine ornamental shrimp *Stenopus hispidus* in captivity. *Aquaculture* 306:185-190.

Haarr ML (2018) *Spatiotemporal variation in sexual maturation and hatching of American lobster (Homarus americanus) in eastern Canada: patterns, processes and implications to fisheries.* PhD thesis, University of New Brunswick, Saint John, NB, Canada.

Hamasaki K, Sugizaki M, Dan S, Kitada S (2009) Effect of temperature on survival and developmental period of coconut crab (*Birgus latro*) larvae reared in the laboratory. *Aquaculture* 292:259-263.

Hamasaki K, Matsuda T, Takano K, Sugizaki M, Murakami Y, Dan S, Kitada S (2016) Thermal adaptations of embryos of six terrestrial hermit crab species. *Aquat Biol* 25:83-96.

Hart RC (1990) Copepod post-embryonic durations: pattern, conformity, and predictability. The realities of isochronal and equiproportional development, and trends in the copepodite-naupliar duration ratio. *Hydrobiologia* 206:175-206.

Hart RC (1998) Copepod equiproportional development: experimental confirmation of its independence of food supply level, and a conceptual model accounting for apparent exceptions. *Hydrobiologia* 380:77-85.

Hartnoll RG, Paul RGK (1982) The embryonic development of attached and isolated eggs of *Carcinus maenas. Int J Inver Rep* 5:247-252.

Hill BJ (1977) The effect of heated effluent on egg production in the estuarine prawn *Upogebia africana* (Ortmann). *J Exp Mar Biol Ecol* 29:291-302.

Jirkowski GJ, Wolff C, Richter S (2015) Evolution of eumalacostracan development – new insights into loss and reacquisition of larval stages revealed by heterochrony analysis. *EvoDevo* 6:4. DOI:10.1186/2041-9139-6-4.

Jirkowski GJ, Wolff C, Richter S (2013) Myogenesis of Malacostraca – the "egg nauplius" concept revisited. *Front Zool* 10:76. DOI: 10.1186/1742-9994-10-76.

Kidd RJ (1991) Development of embryonic and naupliar setae and spines and their role in hatching in the penaeid *Sicyonia ingentis*: a light and electron microscopy study. *J Crustac Biol* 11(1):40-55.

Kitani H, Alvarado N (1982) The larval development of the Pacific brown shrimp *Penaeus californiensis* (Holmes) reared in laboratory. *Bull Jpn Soc Fish* 48(3):375-389.

Kumlu M, Eroldogan OT, Aktas M (2000) Effect of temperature and salinity on larval growth, survival and development of *Penaeus semisulcatus. Aquaculture* 188:167-173.

Kumlu M, Eroldogan OT, Aktas M, Sağlamtimur B (2001) Larval growth, survival and development of *Metapenaeus monoceros* (Fabricius) cultured in different salinities. *Aquac Res* 32:81-86.

Leong PKK, Chu KH, Wong CK (1992) Larval development of *Metapenaeus ensis* (de Haan) (Crustacea: Decapoda: Penaeidae) reared in the laboratory. *J Nat Hist* 26:1283-1304.

MacDiarmid AB, Sainte-Marie B (2006) Reproduction. In: *Lobsters: Biology, Management, Aquaculture and Fisheries (1st Edition)*. Phillips B (ed.). Oxford, UK: Blackwell Publishing, pp. 45-77.

MacKenzie BR (1988) Assessment of temperature effects on interrelationships between stage durations, mortality, and growth in laboratory-reared *Homarus americanus* Milne Edwards larvae. *J Exp Mar Biol Ecol* 116:87-98.

MacNamara KJ (2012) Heterochrony: the evolution of development. *Evo Edu Outreach* 5:203-218.

Martin JW (2014a) Introduction to the Malacostraca. In: *Atlas of Crustacean Larvae*. Martin JW, Olesen J, Høeg JT (eds.). Baltimore, MD, USA: Johns Hopkins University Press, pp. 174-175.

Martin JW (2014b) Introduction to the Decapoda. In: *Atlas of Crustacean Larvae*. Martin JW, Olesen J, Høeg JT (eds.). Baltimore, MD, USA: Johns Hopkins University Press, pp. 230-234.

Martin JW, Davis GE (2001) *An Updated Classification of the Recent Crustacea. Science Series No. 39*. Los Angeles, CA, USA: Natural Museum of Los Angeles County.

Martin JW, Gómez-Gutiérrez J (2014) Euphausiacea. In: *Atlas of Crustacean Larvae*. Martin JW, Olesen J, Høeg JT (eds.). Baltimore, MD, USA: Johns Hopkins University Press, pp. 220-225.

Martin JW, Criales MM, Dos Santos A (2014a) Dendrobranchiata. In: *Atlas of Crustacean Larvae*. Martin JW, Olesen J, Høeg JT (eds.). Baltimore, MD, USA: Johns Hopkins University Press, pp. 235-242.

Martin JW, Olesen J, Høeg JT (2014b) Introduction. In: *Atlas of Crustacean Larvae*. Martin JW, Olesen J, Høeg JT (eds.). Baltimore, MD, USA: Johns Hopkins University Press, pp. 1-7.

Martin JW, Olesen J, Høeg JT (2014c) The crustacean nauplius. In: *Atlas of Crustacean Larvae*. Martin JW, Olesen J, Høeg JT (eds.). Baltimore, MD, USA: Johns Hopkins University Press, pp. 8-16.

Miller RJ (1997) Spatial differences in the productivity of American lobster in Nova Scotia. *Can J Fish Aquat Sci* 54(7):1613–1618.

Miller E, Haarr ML, Rochette R (2016) Using temperature-dependent embryonic growth models to predict time of hatch of American lobster (*Homarus americanus*) in nature. *Can J Fish Aquat Sci* 73(10):1483–1492.

Müller Y, Ammar D, Nazari E (2004) Embryonic development of four species of palaemonid prawns (Crustacea, Decapoda): pre-naupliar, naupliar and post-naupliar periods. *Rev Braz Zool* 21(1):27-32.

Müller YMR, Pacheco C, Simões-Costa MS, Ammar D, Nazari EM (2007) Morphology and chronology of embryonic development in *Macrobranchium acanthurus* (Crustacea, Decapoda). *Invert Rep Dev* 50(2):67-74.

Nagaraj M (1983) Combined effects of temperature and salinity on the zoeal development of the green crab, *Carcinus maenas* (Linnaeus, 1758) (Decapoda: Portunidae). *Scient Mar* 57:1-8.

Newman BK, Papadopoulos I, Vorsatz J, Wooldridge TH (2006) Influence of temperature on the larval development of *Upogebia africana* and *U. capensis* (Decapoda: Thalassinidae: Upogebiidae) in the laboratory. *Mar Ecol Prog Ser* 325:165-180.

O'Connor MI, Bruno JF, Galnes SD, Halpern BS, Lester SE, Kinlan BP, Weiss JM (2007) Temperature control of larval dispersal and the implications for marine ecology, evolution, and conservation. *Proc Nat Acd Sci* 104(4):1266-1271.

Oliphant A, Hauton C, Thatje S (2013) The implications of temperature-mediated plasticity in larval instar number for development within a marine invertebrate, the shrimp *Palaemonetes varians*. *PLoS ONE* 8(9):e75785.

Omori M (1971) Preliminary rearing experiments on the larvae of *Sergestes lucens* (Penaeidia, Natantia, Decapoda). *Mar Biol* 9:228-234.

Omori M (1979) Growth, feeding, and mortality of larval and early postlarval stages of the oceanic shrimp *Sergestes similis* Hansen. *Limnol Oceanogr* 24(2):273-288.

Ouellet P, Chabot D (2005) Rearing *Pandalus borealis* (Krøyer) larvae in the laboratory I. Development and growth at three temperatures. *Mar Biol* 147:869-880.

Pechenik JA (1999) On the advantages and disadvantages of larval stages in benthic marine invertebrate life cycles. *Mar Ecol Prog Ser* 177:269-297.

Perkins HC (1972) Developmental rates at various temperatures of embryos of the northern lobster (*Homarus americanus* Milne-Edwards). *Fish Bull* 70(1):95-99.

Phillips BF, Sastry AN (1980) Larval ecology. In: *The Biology and Management of Lobsters Volume II: Ecology and Management*. Cobb JS, Phillips BF (eds.). New York, NY, USA: Academic Press, pp. 11-57.

Pineda J, Reyns N (2018) Larval transport in the coastal zone: niological and physical processes. In: *Evolutionary Ecology of Marine Invertebrate Larvae*. Carrier TJ, Reitzel AM, Heyland A (eds.). Oxford, UK: Oxford University Press, pp. 141-159.

Poore GCB (2016) The names of the higher taxa of Crustacea Decapoda. *J Crustac Biol* 36:248-255.

Quinn BK (2016) Extra and "intermediate" larval stages in decapod Crustacea: a review of physiological causes and ecological implications, with emphasis on lobsters, *Homarus* spp. In: *Crustaceans: Physiological Characteristics, Evolution and Conservation Strategies*. Alvarado V (ed.). Hauppauge, NY, USA: Nova Science Publishers, Inc., pp. 19-80.

Quinn BK (2018) A test of the general occurrence and predictive utility of equiproportional, isochronal, 'variable proportional', and 'mixed' development among arthropods. Manuscript in preparation to be submitted to *Arthropod Structure & Development* in 2018.

Rao GS (1979) Larval development – *Metapenaeus brevicornis* (H. Milne Edwards). *CMFRI Bull* 28:60-64.

Reisser CMO, Bell JJ, Gardner JPA (2013) Correlation between pelagic larval duration and realised dispersal: long-distance genetic connectivity between northern New Zealand and the Kermadec Islands archipelago. *Mar Biol* 161(2):297-312.

Roberts SD, Dixon CD, Andreacchio L (2012) Temperature dependent larval duration and survival of the western king prawn, *Penaeus (Melicertus) latisulcatus* Kishinouye, from Spencer Gulf, South Australia. *J Exp Mar Biol Ecol* 411:14-22.

Ronquillo JD, Saisho T (1995) Developmental stages of *Trachypenaeus curvirostris* (Stimpson, 1860) (Decapoda, Penaeidae) reared in the laboratory. *Crustaceana* 68(7):833-863.

Ross RM (1981) Laboratory culture and development of *Euphausia pacifica*. *Limnol Oceanogr* 26(2):235-246.

Rötzer MAIN, Haug JT (2015) Larval development of the European lobster and how small heterochronic shifts lead to a more pronounced metamorphosis. *Int J Zool* 2015:345172.

Schneider CA, Rasband WS, Eliceiri KW (2012) NIH Image to ImageJ: 25 years of image analysis. *Nature Methods* 9(7):671-675.

Scholtz G (2000) Evolution of the nauplius stage in malacostracan crustaceans. *J Zool Syst Evol Res* 38:175-187.

Shanks AL (2009) Pelagic larval duration and dispersal distance revisited. *Bio Bull* 216:373-385.

Shokita S (1984) Larval development of *Penaeus (Melicertus) latisulcatus* Kishinouye (Decapoda, Natantia, Penaeidae) reared in the laboratory. *Galaxea* 3:37-55.

Shumway SE, Perkins HC, Schick DF, Stickney AP (1985) Synopsis of biological data on the pink shrimp, *Pandalus borealis* Krøyer, 1838. *FAO Fisheries Synopsis* No. 144.

Silas EG, Muthu MS, Pillai NN, George KV (1979) Larval development – *Penaeus monodon* Fabricius. *CMFRI Bull* 28:2-12.

Stickney AP (1982) Prediction of hatching time for eggs of northern shrimp (*Pandalus borealis*) from measurement of eye pigment spots of the embryos. Unpubl. manuscr., 13 p. *Department of Marine Resources, West Boothbay Harbor*. ME 04575. [Cited in Shumway et al. (1985).]

Talbot P, Helluy S (1995) Reproduction and embryonic development. In: *Biology of the Lobster* Homarus americanus. Factor JR (ed.). San Diego, CA, USA: Academic Press, pp. 177-216.

Tamaki A, Tanoue H, Itoh J, Fukuda Y (1996) Brooding and larval developmental periods of the callianassid ghost shrimp, *Callianassa japonica* (Decapoda: Thalassinidea). *J Mar Biol Assoc UK* 76:675-689.

Taveras C, Martin JW (2010) Suborder Dendrobranchiata. In: *Treatise on Zoology, the Crustacea. Vol. 9A*. Schram F, von Vaupel Klein JC (eds.). Leiden, pp. 99-164.

Thomas MM, Kathirvel M, Pillai NN (1974) Spawning and rearing of the penaeid prawn, *Metapenaeus affinis* (H. Milne Edwards) in the laboratory. *Ind J Fish* 21(2):543-556.

Villarreal H, Hernandez-Llamas A (2005) Influence of temperature on larval development of Pacific brown shrimp *Farfantepenaeus californiensis*. *Aquaculture* 249:257-263.

Wahle RA, Tshudy D, Cobb JS, Factor J, Jaini M (2012) Infraorder Astacidea Latreille, 1802 P.P.: the marine clawed lobsters. In: *Treatise on Zoology, the Crustacea. Vol. 9B*. Schram FR, von Vaupel Klein JC (eds.). Leiden, pp. 3-108.

Williamson DI (1982) Larval morphology and diversity. In: *The Biology of the Crustacea, Vol. 2: Embryology, Morphology, and Genetics*. Abele LG (ed.). New York, NY, USA: Academic Press, pp. 43-110.

Williamson DI (1992) *Larvae and Evolution: Toward a New Zoology*. New York, NY, USA: Chapman & Hall.

Williamson DI (2003) *The Origins of Larvae*. Dordrecht, The Netherlands: Kluwer.

Williamson DI (2006) The origins of crustacean larvae. In: *Treatise on Zoology, the Crustacea. Vol. 2*. Forest J, von Vaupel Klein JC (eds.). Leiden, pp. 461-482.

WoRMS Editorial Board (2018) *World Register of Marine Species*. Available from http://www.marinespecies.org at VLIZ. Accessed 2018-06-20. DOI: 10.14284/170.

Yamamoto T, Jinbo T, Hamasaki K (2017) Intrinsic optimum temperature for the development of decapod crustacean larvae based on a thermodynamic model. *J Crustac Biol* 37:272-277.

Ziegler TA, Forward RB, Jr. (2007) Control of larval release in the Caribbean spiny lobster, *Panulirus argus*: role of chemical cues. *Mar Biol* 152:589-597.

In: Focus on Arthropods Research   ISBN: 978-1-53614-343-0
Editor: Mirko Messana   © 2018 Nova Science Publishers, Inc.

Chapter 3

ARTHROPODS AND THEIR PATHOGENS: THE EPIZOOTIOLOGY AND BIOLOGY OF SOME GREGARINES (PROTOZOA: APICOMPLEXA) FROM *SCAURUS PUNCTATUS* (COLEOPTERA: TENEBRIONIDAE), WITH DATA ON THEIR IMPACT CONCERNING INSECT FITNESS

A. Criado-Fornelio, C. Verdú-Expósito,
T. Martín-Pérez, I. Heredero-Bermejo
and J. Pérez-Serrano*
Laboratorio de Parasitología, Departamento de Biomedicina y
Biotecnología, Facultad de Farmacia, Universidad de Alcalá,
Alcalá de Henares, Madrid, Spain

---

* Corresponding Author Email: angel.criado@uah.es.

ABSTRACT

Some arthropods, like the beetle *Scaurus punctatus* (Coleoptera: Tenebrionidae) are considered benefitial in terrestrial ecosystems of southern Europe, due to their detritivore activity. There is little information available on their parasites and the effect they cause on beetles. This chapter provides a new insight on aspects such as epizootiology, biology and impact of gregarines in these insects.

Epizootiological studies in this laboratory have led to the first finding of *Gregarina wharmani* and *Gregarina* sp (n°2) in *S. punctatus* beetles from Alcalá de Henares (Madrid, Spain). Since these organisms were described first in Israel and Maroc, this report represents the first finding in Europe. Interestingly, fully developed gamonts of *G. wharmani* show a sturdy pellicle or epicyte (a superficial layer) that stands the passage through the insect's intestinal tract. Therefore, this gregarine stage can be diagnosed by microscopical examination of feces.

A previous study conducted in our laboratory has described in detail the morphology of *G. ormierei*. As a continuation of such investigations, some new aspects of the biology and morphology of *G. ormierei* are presented here:

1. Sporozoite release from the oocyst relies on a mechanism related to that seen in some Ciliates.
2. In the small phenotype, a cytoplasmic tail anchors the gamonts in syzygy to the insect's peritrhropic membrane,
3. Undehyscent gametocysts of both phenotypes are many times mature and contain viable oocysts capable of infecting insects,
4. Trophozoites show first a globular epimerite that later becomes conical/tubular. Such a new evidence suggests that *G. ormierei* and *G. cavalierina* might be the same organism. Histological examination of beetle gut tissue is neccesary to confirm such hypothesis.

Diverse environmental factors, like temperature and humidity, affect the populations of *G. ormierei*. Gregarine prevalence in beetles reaches a peak in late spring and decreases in summer. The life cycle duration is 10-25 days, depending on gregarine phenotype. Finally, to monitor possible detrimental effects of *G. ormeirei* on beetles, protozoan load and different morpho-biological characteristics of naturally infected and uninfected insects were analysed for correlation. There was significant negative correlation ($P < 0.05$) between gregarine load and beetle weight, body length and abdomen width, whereas positive correlation existed with both bold behavior and amputations ($P < 0.05$). The present results suggest that *G. ormierei* causes harmful effects in *S. punctatus*. Such finding is ecologically meaningful, since this abundant insect is involved in the natural transformation of decaying organic material in soil.

Keywords: *Scaurus punctatus*, Coleoptera, gregarines, epizootiology, biology, detrimental effects

INTRODUCTION

In Southern Europe and North Africa, beetles such as *Scaurus* spp (Coleoptera: Tenebrionidae) play a beneficial role in nature. These abundant arthropods develop a significant detritivore activity, contributing to decompose both plant and animal biomaterial in soil (Cartagena and Galante, 2003). But despite their important ecological function, current knowledge on pathogens of Tenebrionid beetles is limited (Clopton, 2009; Desportes and Schrevel, 2013, Criado-Fornelio et al., 2013 and 2017). In this regard, gregarines (Protozoa: Apicomplexa) are one of the most common intestinal parasites found in insects. In the Mediterranean area (particularly in France, Italy, Morocco and Israel), gregarines infecting *Scaurus* spp were described or revised by Filipponi (1952a) and Theodorides (1952 and 1955a, b, c), who reported only two species:

1. *Gregarina cavalierina*, originally described by Blanchard (1905) in *Dendaurus tristis* in France and subsequently redescribed by Theodorides (1952) in *Scaurus striatus* and *S. gigas*, based on their own observations and also on those by Filipponi (1952a) in Italy. Interestingly *G. ormierei*, a gregarine apparently related *to G. cavalierina*, was incompletely described (gametocyst and oocyst morphology was not published for this species) in other tenebrionid beetles in France, Turkey, Sudan, Ethiopia and Korea (Theodorides, 1955a and 1958a; Theodorides et al., 1964, 1965 and 1976).

2. *G. wharmani*, described by Theodorides (1955c) in *Scaurus puncticollis* in Israel. The same as for *G. ormierei*, both gametocyst and oocyst stages are unknown.

In Spain, Theodorides (1958b and 1960) performed a survey for gregarines in beetles in the Canary Islands and Murcia, respectively. Unfortunately, *Scaurus* spp were not sampled in these former works. A study conducted in Alcalá de Henares by our own research group failed to detect gregarines in a small sample of *S. punctatus* by molecular methods (Criado-Fornelio et al., 2013). Nevertheless, morphological studies recently published by our research team (Criado-Fornelio et al., 2017) led to the redescription of *G. ormierei* in these beetles. Due to the paucity of data available on protozoal fauna of insects in Spain, the present work is an attempt to deepen our current knowledge on gregarine epizootiology and biology. The aim of the study is the morpho-biological characterization of gregarines found in *S. punctatus* Fabricius 1789 in Alcalá de Henares (Madrid, Spain). In addition, the possible negative effects of gregarines on beetle fitness were analysed, an issue often neglected in former contributions dealing with parasitic protozoa of Coleoptera in Europe.

MATERIAL AND METHODS

Field Sampling for Epizootiological or Morpho-Biological Studies

Beetles (*Scaurus punctatus* Fabricius 1789; Coleoptera, Tenebrionidae) were captured in an uncultured land near to the Facultad de Farmacia at the Universidad de Alcalá in Alcalá de Henares, Madrid, Spain (approximate geographic coordinates: 40° 30' 35" N; 3° 20' 27'' W). Specimens were obtained either by hand or pitfall traps, as described by Greenslade (1964). Identification of *Scaurus* spp was based on Español (1968). One hundred and fifty beetles were destined to epizootiological studies, while nearly 400 specimens were employed in studies on gregarine biology and their possible deletereous effects on insects. The beetle *S. punctatus* is not a protected species under the Spanish law, so that there were no issues related to Nature conservation.

Laboratory Maintenance of Beetles for Recovery of Gametocysts and Oocysts

Scaurus punctatus beetles were individually maintained in 8 cm diameter Petri dishes and fed with small apple pieces and fragments of decaying leaves of *Platanus hispanica* trees (the latter were collected from the same field mentioned above, so that they were potentially contaminated with gregarine oocysts). Insects were occasionally fed with small portions of preserved pork meat (Spam® or similar), to increase the protein content of the food. Insects were kept in a controlled environmental chamber (Ibercex C-3, Madrid, Spain) at 24°C under a light-dark schedule of 12:12. All Petri dishes were examined daily under stereo microscope. Gametocysts were washed with sterile insect saline solution (1.8% NaCl, 1.88% KCl, 0.16% $CaCl_2$ and 0.004% $NaHCO_3$ w/v in distilled water). They were loaded with a micropippette in a concave microscope glass slide. They were maintained at 24°C either in air (slides were placed in a wet chamber) or alternatively in saline (the concave slide was filled up with saline solution, covered with a coverslip and placed in a wet chamber). Gametocyst status was observed daily under microscope to record dehiscence time.

Microscopic Studies

Insects were sacrificed and the gut was removed for analysis. The intestine was deposited in a microscopic glass slide and it was teared in a drop of insect saline solution. This facilitated microscopic observation of the intestinal content preserving gregarine integrity. Morphologic features were measured in microphotographs obtained in a Motic BA300 microscope equipped with a digital camera and employing the Motic Images Plus software version 2.0. Morphometric measurements obtained in the present work follow terminology proposed by Clopton (2004) and Janovy et al., (2007), and were made on live specimens (whenever feasible) as recommended by Clopton et al., (1992). Sporozoite nuclei in oocysts were stained with Ecoline 111 (Royal Talens, Apeldoorn, The Netherlands). Stained oocysts were mounted in Hoyer's medium as described by Moreno and Manjon (2010). Biological nomenclature of gregarine stages follow the terminology by Levine (1971). Due to the low frequency of infection, measurements (in micrometers) are the result of single observations, in contrast to a previous study conducted by us in *G. ormierei* (Criado-Fornelio et al., 2017).

In scanning electron microscopy, gregarine sporozoites were directly observed (without fixation and dehidration in acetone series) as recommended by Janovy et al., (2007). Gametocysts (and the oocysts they contain) were squashed under two glass coverslips, fixed and processed for observation of sporozoites. Samples were sputter-coated with 200 Å gold-palladium using a Polaron E5400. Scanning electron microscopy was performed at 5-20 kV in a Zeiss DSM 950 SEM.

Experimental Infection of Beetles

After dehiscence of gametocysts in air, oocysts were recovered from concave slides by rubbing a small piece of apple against the glass slide in the area of dehiscence. Due to both the porous structure of apple flesh and the presence of juices, most oocysts on the microscopic slide adhered to the apple fragment. This was assessed by microscopical examination of the concave slide after rubbing. Big gametocysts contained approximately 3000-4000 oocysts, so that this was the dose ingested by one beetle after spontaneously eating the contaminated apple portion deposited in its individual Petri dish. For indehiscent gametocysts and small gametocysts (which usually were enclosed in insect's peritrophic membrane), they were rehydrated with saline solution and recovered with either a pippete or extrafine forceps under a stereomicroscope and then transferred into a small fragment of a sterile *P. hispanica* decaying leaf where they were left for drying. The leaf fragment was used as food for one beetle.

Beetles employed in experimental infections were previously deprived of gregarines. This was achieved by keeping them at 37°C for 9 days, as explained below under the "results" section.

Excystation Test

For excystation tests, a minimum of twelve *S. punctatus* were sacrificed. Insect gut was removed, cut in small portions, transferred into a 1.5 ml Eppendorf tube and centrifuged at 10.000 x *g* for 5 min at 4°C. Fifty µl of supernatant (intestinal fluid) were added to a big gametocyst (which had suffered dehiscence on a microscopic glass slide). Therefore, a great number of oocysts chains had been adhered to the glass surface. A coverslip was placed on the preparation, which was sealed with nail varnish and incubated in a wet chamber at 24°C. Observations were made under light microscope every 15 min until excystation was noticed.

Studies on the Effect of Gregarine Load on Insect Morphology and Biological Fitness

Scaurus punctatus beetles were employed as model system to study the effect of gregarine infection on morphological and biological features of insects. Fifty beetles (25 male and 25 female) were captured in the field at the end of May (when most of them were naturally infected with gregarines) and individually maintained in Petri dishes at 24°C. Beetles were sacrifized after one week. Food consisted in decaying leaves of *P. hispanica* (+ Spam®) collected in the capture site (to ensure that infection persisted in beetles) and was administered *ad libitum*. The following parameters were monitored:

(a) Fecal production. The feces produced by a beetle in a week were recovered from the Petri dish with the help of a fine brush. Fecal material drying was done by incubation at 50°C overnight. Once dry, feces were weighed. The fecal production was calculated as a ratio in relation to body weight (FP = feces weight per week/beetle weight).

(b) Response to light/dark. On the seventh day, insects were individually transferred from their usual Petri dish with food to another Petri dish (without food) which was half covered by a black cardboard and illuminated with a 25 W incandescent bulb so as to produce a shaded and a lighted area. Insects were placed in the interface between light and dark. The observer hid from insect sight and after waiting for 15 seconds animal's behaviour was observed and recorded. Three possible results were considered: Neutral (the insect did not move - code for statistical analysis = 1); displacement to shaded area (code = 2) and displacement to lighted area (code = 3).

(c) Response to threat. The experiment was designed to detect incautious behaviour or alternatively, thanatosis (the latter, also called death-feigning, is a state of immobility assumed by many insects in response to external stimuli, like predators). At seventh day, beetles were individually caught with forceps from their Petri dish and deposited upside down in the laboratory desk in presence of the observer, who repeatedly hit the desk surface with a stick at about 15 cm of the insect (as a mimick of threat). Response (after waiting for 15 seconds) was classified in two categories: Death-feigning, when the insect lies upside down without moving (code = 1) or bold behaviour, when the insect gets back to the upright position and either stays still or escapes (code = 2).

Once behavioural studies were finished, beetles were euthanized with chlorophorm and some additional parameters were monitored: insect weight, total body length, maximum abdomen width, amputations (either limbs or antennae). Beetle sex (which had been recorded at the beginning of the experiment) was coded with 1 (male) or 2 (female) for statistical purposes. After necropsy, the type and number of gregarines present in beetles gut was studied. Microscopic slides marked with grid were employed to facilitate parasite counting at 40X under stereomicroscope.

Studies on the Effect of Gregarine Load on Female Reproductive Capacity

Egg production in female beetles was analysed as follows: 20 couples (male and female beetles) were placed in 20 Petri dishes. Animals were left in an incubator at 24°C for 30 days. During this period, insects were fed *ad libitum* with decaying leaves (+ Spam®) from the capture site. Every day, the eggs laid by female beetles withdrawn from the dish and counted. The monthly egg production was recorded at the end of the experiment. At this moment, all live females were sacrificed, weighed and their gregarine load counted. The egg production was calculated as a ratio in relation to body weight (EP = eggs per month/female beetle weight).

Monthly Variation of Gregarine Load and Phenotype

From April until the end of July and also in September, ten beetles were captured in the field every week and examined for gregarine infection. Abundance for the two *G. ormierei* phenotypes present in beetles and prevalence (for both phenotypes) was obtained every month (except in August, due to vacation of personnel performing field work). Records of average monthly temperature and rainfall for Alcalá de Henares (Anonimous, 2008) were employed (along with parasitological data) in combined curve plots.

Effect of Temperature on Gregarine Phenotype

Three experiments were designed in order to study the effect of temperature on gregarines:

Experiment 1 (survival of trophozoites at low temperature): one group of six beetles was kept for one month at 4°C. All the insects were killed after the experimental period and intestinal gregarines observed and counted. No food was provided to the animals, as they were paralysed by cold.

Experiment 2 (development of trophozoites at moderate temperatures): one group of 28 beetles that had been in the laboratory for one month at 24°C before the beginning of the experiment was divided in two groups (14 individuals each); one group was kept at 24°C and the other at 32°C. All the insects in both groups were killed after 15 days and intestinal gregarines counted. In this case *Platanus* leaves (+ Spam®) from the capture site were used to feed the animals.

Experiment 3 (development of trophozoites at extreme temperature): one group of 18 beetles was kept at 37°C and three samples of six beetles each were examined for gregarines at days 3, 6 and 9. In this case sterile *Platanus* leaves (+ Spam®) were used to feed the animals, to avoid reinfection.

Statistical Analysis

Variable correlation analysis was performed with either the STATGRAPHICS PLUS® (Statpoint Technologies Inc., Warrenton, VA, USA) or GRAPHPAD PRISM 5® (GraphPad Software, San Diego CA, USA) software packages. Statistical significance was set at $P < 0.05$.

RESULTS

Epizootiology of Gregarines in *S. punctatus*

Two gregarines (apart of *G. ormierei*) were found in *S. punctatus*. Gamonts of *G. wharmani* (Figure 1A) and *Gregarina* sp n°2 (Figure 1C) were observed in the intestine of two different beetles. Therefore, both parasites showed 0.66% prevalence (1 infected/150 studied). Microscopic observation of beetle feces (performed when searching for gametocysts of the small phenotype gregarine in other biological studies) unequivocally identified another dead gamont of *G. wharmani* (Figure 1B). Since this observation was made in beetles not included in the epizootiological survey, it was not taken into account for prevalence calculation.

Measurements taken in single protozoa specimens showed than the live gamont of *G. wharmani* was 150 µm long (deutomerite lenght was 125 µm and protomerite length was 25 µm). The nucleus diameter was 20 µm. The dead specimen of *G. wharmani* (detected in feces, Figure 1B) was bigger: 460 µm (deutomerite length: 385 µm; protomerite length: 75 µm). The nucleus diameter was 50 µm. The protomerite in this species was shallowly to very broadly ovoid, in contrast to the obpanduriform shape seen in gamonts of *G. ormierei*. On the other hand, the gamont of *Gregarina* sp. n° 2 measured 60 µm (with deutomerite length 55 µm and protomerite 5 µm long). The apical part of the rectangular protomerite was slightly granulous inside and with wrinkled surface, very different from that of either *G. ormierei* or *G. wharmani*. No post-nuclear stain was apparent.

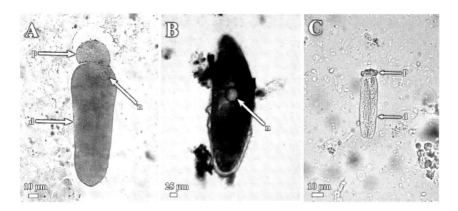

Figure 1. Light microscopy micrographs showing different stages of two gregarines found in the epizootiological survey conducted in *Scaurus punctatus*. All micrographs: p = protomerite; d = deutomerite, n = nucleus. A. Gamont of *Gregarina wharmani* in *S. punctatus* gut (stained with methylene blue in permanent mount). Cytoplasmic granules are quite transparent compared to those of *G. ormierei*. B. Dead gamont of *G. wharmani* in *S. punctatus* feces (wet mount). C. Gamont of *Gregarina* sp. n°2 in *S. punctatus* gut (wet mount).

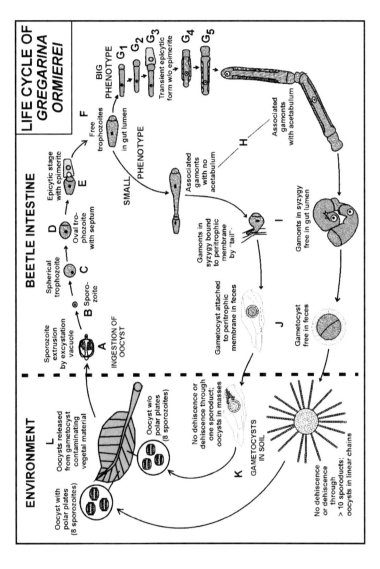

Figure 2. Diagrammatic representation of *G. ormierei* life cycle. Stages "A" to "F" are common for both phenotypes. A dichotomy occurs in stage "F", where gregarines split into either big or small phenotype. Stages "G_1" to "G_5"are immature trophozoites/gamonts of the big phenotype that may be solitary in gut lumen or in some periods (e.g., stage G_4) may be attached to insect intestinal cells.

Biology of *G. ormierei* in *S. punctatus*

A scheme of the life cycle of *G. ormierei* in *S. punctatus* is depicted in Figure 2, and details of its different biological stages are shown in Figures 3, 4 and 5. Oocysts were polizoic, with 8 sporozoites measuring 1.2 to 1.5 μm in diameter (Figure 3A). Some intact oocysts of this species showed an inner axial tubular structure going from one pole to another (Figure 3B). It measured 0.5 μm in diameter and about 5-6 μm in length. Other oocysts presented a great vacuole at one side (Figure 3C). Such structures, however, were not evident in all oocysts. When sporozoite excystment was analysed *in vitro*, the function of both the oocyst vacuole and axial tubule become clear. Sporozoite emergence started after 3 h of incubation of gregarine oocysts in *S. punctatus* intestinal juice at 24°C. When sporozoite extrusion started, the distal part of the axial tubule protruded about 1 μm out of the oocyst wall, first at one pole where the sporozoite was ready to start excystation (Figure 3D) and later at both poles (Figure 3E). The extrusion of the tubule and the sporozoite seemed to be due to the pressure created by the excystment vacuole that gradually swelled inside the oocyst. Once the vacuole had swollen in one side of the oocyst, this part appeared clearer than the rest, which suggest that there was passive diffusion of water from the environment into the cell (Figure 3E). Finally, the pressure caused by the vacuole in one side causes the rupture the blunt end of the conduct at the opposite end. This probably debilitates the oocyst shell as well. Sporozoites are located in the space existing outside the axial tubule, but they seem capable to locate the hole in the wall and push through the opening in the excystation process. Activated oocysts in extrusion phase were floating in wet mount preparations, which suggests that their density had changed respect to quiescent ones. Such flotability may be due to the intake of water from the external medium mentioned above. Sporozoite extrusion was relatively slow. The cell struggled to escape, due to the fact that protozoa diameter seemed to be slightly greater than the orifice opened by the rupture of the oocyst wall. Sporozoite emergence was finished after more than 3 minutes of continuous contractions of the protozoa at the narrow opening of the oocyst wall hole. The process of oocyst hatching occurs when it is floating in medium, and the sporozoite is shaking in its attempt to exit through the shell; this is why pictures (obtained by optical microscopy with immersion oil) are not in a very sharp focus. However, our direct observations were clear enough as to ensure that the hatching process occurred as depicted here. Once the sporozoite was free, it was capable of moving at low speed, although no flagellum was observed at this stage. Probably such displacement was caused by gliding motility.

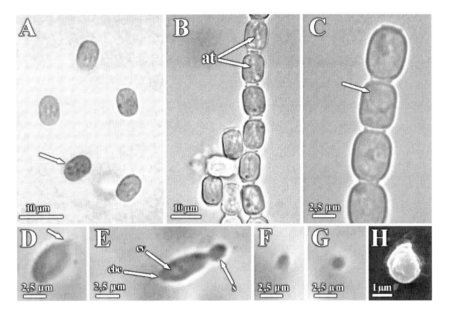

Figure 3. Microscopic observations of different stages of *G. ormierei*. A. Oocysts of the big genotype stained with Ecoline 311 in Hoyer's medium. Up to eight sporozoites per oocyst may be observed (arrow). B. Big phenotype oocyst chain; arrows with "at" indicate an axial tubular structure inside the oocyst. C. Oocyst chain (big phenotype). One of the oocysts contains a big circular vacuole (arrow). D. Oocyst incubated in *S. punctatus* intestinal juices for 3 h. The arrow indicates a polar protuberance where a sporozoite is starting its extrusion process through the axial tubule. E. The same cell in 3D at 3h 3 min of incubation. Arrow with "s" indicates the sporozoite performing contractions, pushing for excystment. Arrow with "ebe" indicates the presence of an emergent blunt end at the opposite pole; in that end a relatively clear space (the excystment vacuole), may be seen inside the oocyst (arrow with "ev"). F. Motile oval sporozoite emitted from the oocyst seen in 3D,E at 3 h 8 min. G. Motile spherical sporozoite at 3 h 10 min. H. SEM micrograph of a *G. ormierei* sporozoite.

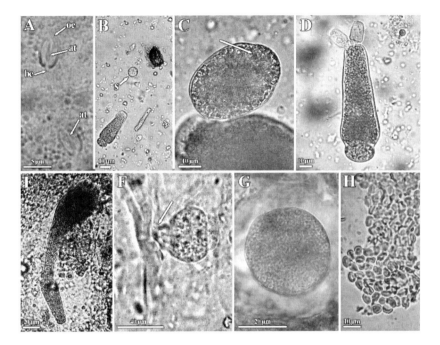

Figure 4. A. Oocyst shell after excystment. The arrows with "at" points to the axial tubule; arrow with "be" indicates the unopened blunt end of the axial tubule and arrow with "oe" denotes the open end (pole) of the oocyst. B. A small trophozoite with spherical shape present in fecal material (arrow with "s"). C. Oval trophozoite with evidence of apparition of septum (arrow). D. Triple association (small phenotype). E. Association with primite deutomerite becoming oval at the septum region and elongated at the rear. F. Dead mature gamont association with a "tail" (arrow) embedded in peritrophic membrane. To obtain this image, dry beetle feces were resuspended in insect saline. Gamonts of *G. ormierei* were very difficult to find in dry beetle feces, therefore this specimen was a very rare exception. G. Small gametocyst in a wet mount preparation of beetle intestine. No oocysts are visible yet. H. Squash preparation of an undehiscent small gametocyst recovered in beetle feces. Oocysts are present in the scattered gametocyst content.

Sporozoites keep their oval shape for a couple of minutes after extrusion (Figure 3F), but it became spherical soon afterwards (Figure 3G). Sporozoites showed a relatively smooth surface when observed under SEM (Figure 3H). After hatching, empty oocysts clearly showed the axial tubule, intact at one side (that near to the excystation vacuole, albeit the latter was not visible at this stage) and open at the other pole (Figure 4A). The hole in the oocyst wall is of approximately one μm in diameter, smaller than the size of sporozoites. This explains why the protozoa must struggle to hatch.

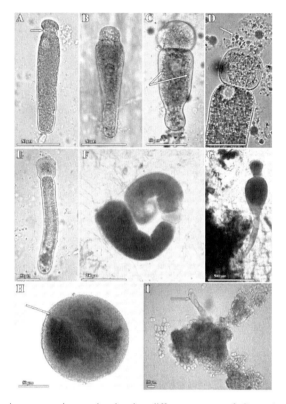

Figure 5. Light microscopy micrographs showing different aspects of *G. ormierei* life cycle (big phenotype). Part 1: Transformation from small to big phenotype (phases G1 to G5 in Figure 2): A. Young gamont where the protomerite is starting to elongate from the septum base (arrow). Deutomerite is still flat-ended. B. Young gamont with protomerite in elongation, vesicular nucleus and deutomerite with blunt end. C. Young gamont with protomerite almost orbicular and deutomerite with a bulky region where a sarcocyte layer has started to develop (arrows). D. Young trophozoite showing cellular debris in protomerite apex (arrow), with no evidence of epimerite. The nucleus with its nucleolus is in the anterior part of deutomerite. A lateral sarcocyte layer is also visible. E. Young gamont showing most of the cardinal features of the big phenotype: protomerite obpanduriform or broadly oblong and blunt-ended deutomerite with a sarcocyte layer. Probably due to a relatively rapid growth period, the dark granulation typical of mature trophozoites is not present yet. Part 2: Syzygy and gametocyst. F. Gamonts in standard syzygy, associated through the twisted acetabulum and standing side by side. G. Gamonts in syzygy; the association shows a "tail". This life cycle stage resembles that found in the small phenotype. Note that the protomerite is obpanduriform (different to the shallowly to broadly ovoid seen in the small phenotype). Primite deutomerite is broadly obovoid and the rear part of the association is in elongation. The tail probably penetrates in the peritrophic membrane, although this part it is not visible due to cellular debris present in the wet mount preparation. H. Carmine stain of a fully formed gametocyst recovered from beetle gut content. A junction line between both gametocytes is still visible (arrow), but there are no oocysts inside. A transparent mucous layer covered the gametocsyt but it was lost during the fixation and staining process. I. Squash preparation of an undehiscent big gametocyst. Normal oocysts with polar plates and a short sporoduct with oocysts inside (arrow) can be observed.

It is likely that the sporozoites released from the oocyst became soon intracellular, living in insect intestinal cells. However, such hypothesis is merely speculative, since we did not prepare histological sections of intestinal tissue in this study. In experimental infections of *S. punctatus* beetles, spherical trophozoites that had grown up to 10 µm in diameter were present in insect gut at 24 - 48 h post infection (Figure 4B). Both septum and epimerite were not apparent in this gregarine stage, but later trophozoites started to adopt an oval shape (Figure 4C) and soon both structures were present in many of them. It is unknown if oval trophozoites were intracellular or not, although probably they are too big as to accomodate inside one insect gut cell. On the third day p.i., immature associations were numerous. Triplets consisting in precocious asociations of one primite and two satellites ("clover-shaped" configuration) were present only in heavy infections (Figure 4D).

The number of associated gamonts steadily increased in days 3 – 6 p.i., but there were still many individual gamonts in gut lumen. On the 5th day p.i., many associations showed a curious morphological change: the rear end of the primite deutomerite begun to elongate, whilst its anterior region became ellipsoid (Figure 4F). The function of the elongated region or "cytoplasmic tail" of the association was unknown and could be unveiled only when insect's peritrophic membranes were examined in beetle feces. It was evident that some dead gamonts showed the cytoplasmic tail embedded in the peritrophic membrane, thus helping to ensure an optimal attachment for the future gametocyst (Figure 4F). Such preservation of associations was very rare, as usually they ended either destroyed in gut or evolved into gametocysts. Small gametocysts observed in gut squash preparations contained no oocysts (Figure 4G). Gametocysts appeared in fecal material from day 10 p.i., but due to the difficulties inherent to searching for these small structures in peritrophic membranes, it is possible that their emission could have started even earlier. It is important to underline that when peritrophic membranes containing indehiscent small gametocysts were recovered from feces and subject to squash for microscopic observation, most ruptured gametocysts contained apparently normal oocysts (Figure 4H).

The life cycle in the big phenotype was identical to that of the small form in the first stages, with spherical, oval and jug-shaped phases, the latter with epimerite (Figure 2, stages "A" to "F"). Once the epicytic period ended, there was evidence of switching from small to big phenotype. Changes consisted in extension in length of the protomerite in gregarines similar in size (or slightly longer) than primite gamonts of the small phenotype (Figure 5A). In a further step, the nucleus changed from compact to vesicular and the flat deutomerite end changed to a blunt shape (Figure 5B). Later, a sarcocyte layer developed in the deutomerite, starting in its anterior end, which appeared slightly swollen (Figure 5C). At this moment, protomerite adopted orbicular or broadly oblong shape and trophozoites were capable of direct attachment to gut cells through the surface of the protomerite apex (Figure 5D). Thus, it seems that gamonts of the big phenotype need some extra feeding on host intestinal cells due to their larger size. In a later stage, deutomerite elongates and protomerite may be broadly oblong to broadly obpanduriform (Figure 5E). In the next life cycle step, gamonts in gut lumen grew up, reaching between 400 to 900 µm in length, approximately. Caudo-frontal gamont associations with a crescentic interface started to appear later (Figure 5F); in this case, no gamont in triplets were present in the gut content, in contrast with the small phenotype. Gamonts in syzygy showed a broadly oblong deutomerite, and bended upon themselves resulting in a head-tail arrangement (Figure 5F). Exceptionally, associations in syzygy of the big phenotype were bound to the peritrophic membrane through a "tail" (the same as in the small phenotype - Figure 5G). Free gametocysts recovered in insect gut still contained two hemispherical gametocytes but no oocysts (Figure 5H). Gametocysts started to appear in beetle feces approximately at day 25 p.i. Gametocyst dehiscence by sporoducts occurred within a period of 3-5 days in air and 2 – 4 days in saline. In the big phenotype gametocysts, dehiscence showed a frequency of 41.66% (10/24) in saline and only 15% (21/122) in air. To investigate the possible maturation and survival of gametocysts without evident dehiscence and outer sporoduct formation, 50 dry and apparently unripe gametocysts recovered from feces were rehydrated in insect saline solution and then placed (with a micropippete) on a microscopic glass slide. These were squashed with a coverslip for microscopic observation of their content. Seventy-six % (43/50) of them harboured apparently normal oocysts. Some of those gametocysts even contained short sporoducts (Figure 5I). Fourteen percent of the indehiscent gametocysts (7/50) contained just an amorphous granular mass with no oocysts at all. In experimental infection assays with 20 undeshicent gametocysts, 60% (12/20) produced successful experimental infection in beetles. Only 3.3% (1/30) of the small gametocysts observed in feces were dehiscent.

It is important to underline that when oocysts of the big phenotype were administered to gregarine-free beetles (at 24°C), most trophozoites and gametocysts later detected in gut or feces were of the small phenotype. This suggests that temperature affects phenotype expression. In fact, small phenotype gregarines derived of oocysts of the big phenotype in laboratory infections. In addition, data on occurrence of both phenotypes in gregarine populations obtained in nature in different periods of the year (in the spring and summer season) seem to support such hypothesis (see section below).

Observations on Epimerite Morphology in Trophozoites of *G. ormierei*

Monitorization of the experimental infections conducted in *G. ormierei* showed that the epimerite was not only orbicular, but also appeared as either oblong (cylindrical) or conical/tubular (Figure 6, A to D).

Monthly Variation of *G. ormierei* Infection Rate in *S. punctatus*

Population dynamics of gregarines (abundance and prevalence), along with climatic data (monthly average temperature and rainfall) are shown in Figure 7. Small phenotype gregarines showed high abundance in April with gradual decrease in the following months. In contrast, big phenotype gregarines were rare in middle-spring, reached a peak at the beginning of summer and then reduced their parasitization levels in parallel with the small type. Gregarine prevalence increased from April to June, whereas it decreased starting in July (in coincidence with low rainfall and high temperature values).

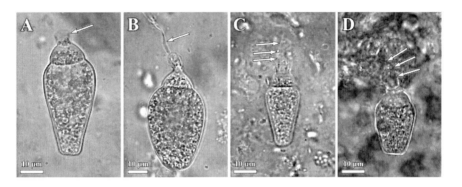

Figure 6. Light microscopy micrographs showing different aspects of *G. ormierei* epimerite in experimental infections performed in *S. punctatus*. Arrows point to the epimerite. A. Cylindrical epimerite. B. Oblong-tubular epimerite, C and D. Conical epimerite. D.

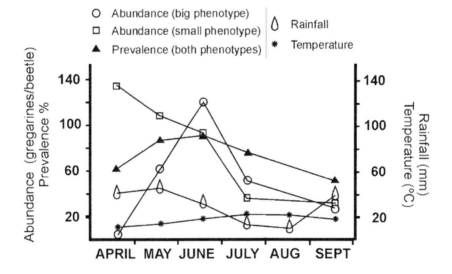

Figure 7. Plot of monthly variation (April to September) of average temperature, rainfall and parasitological variables related to *G. ormierei* infection in *S. punctatus* beetles in Alcalá de Henares.

The Effect of Temperature on *G. ormierei* Development

In experiment 1 (survival of trophozoites at low temperature), all beetles were alive at the end of the experiment. Examinations of gut content revealed that 5 insects were infected with both small and big phenotype gregarines. The small gregarines were still alive since some of them started to show gliding movement when they were in the heat of the microscope lamp. In contrast, big gregarines were motionless and their cytoplasmic granulation was so clear that both nucleus and nucleolus could be observed even without carmine staining (Figure 8).

In experiment 2 (development of trophozoites at moderate temperatures), the mean number of small gregarines/beetle was 238.28 $\pm$ 517.56 at 24°C and 77.71 $\pm$ 129.27 at 32°C (such difference was non-significant). The same parameters for big gregarines were 0.5 $\pm$ 0.85 at 24°C and 39.57 $\pm$ 132.66 at 32°C ($P <$ 0.007 by the Kolmogorov-Smirnov test).

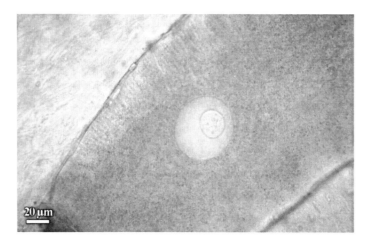

Figure 8. Detail of the unstained deutomerite of a big phenotype gamont of *G. ormierei* found in the gut of a beetle kept at 4°C for one month. Both the nucleus and the nucleolus are evident, due to the great transparency of cytoplasmic granulation. The existence of some alterations in the epycite suggests that the motionless protozoa might be dead.

In experiment 3 (development of trophozoites at extreme temperature – 37°C), two of the beetles (out of six) examined at the third day were infected;

one of them was parasitized by small phenotype gregarines (24 protozoa) and another by big phenotype gregarines (7 protozoa). At day 6, another two beetles out of six were infected; one by two and the other by four big gregarines, respectively. The remaining 6 beetles, examined at day 9, were gregarine-free. Therefore, maintenance of insects for 9 days at 37°C was a possible method to eliminate gregarines of beetles.

Study on the Effect of *G. ormierei* Load on Beetle Morphology and Biology

Table 1 shows the results obtained in the study carried out in a group of 50 beetles. The variables weight, length and abdomen width were significantly lower in infected beetles ($P < 0.05$). The number of amputated limbs and the display of bold behavior was greater in infected beetles ($P < 0.05$). Analysis of correlation for other variables (beetle sex, response to light and fecal production) revealed no association with gregarine load.

Multivariable analysis led to the best fitting model employing the Poisson regression, with the following equation:

$$NG = \exp (5.89 - 5.9*TW - 2.78*TL + 0.53*AM + 0.52*THR)$$

where NG = No. of gregarines; TW = Total weight; TL = Total Length; AM = No. of amputations and THR = response to threat.

The intensity of gregarine infection was negatively associated with weight and total length, whereas the variables "amputations" and "response to threat" showed positive correlation (Chi square, $P < 0.0001$). The predictive value of the Poisson model is limited, since $R^2 = 0.335$. In summary, this model indicates that beetles with low weight, reduced size, bold behaviour and amputations were more likely to harbour gregarines. This is quite similar to the results obtained by paired analysis of variables, although in the Poisson model "abdomen width" was rather a low weight variable.

Table 1. Comparison of diverse morphological, biological and behavior parameters between *G. ormierei*-infected and uninfected *S. punctatus* beetles (N = 50, 25 males + 25 females)

STATUS (right)/Variables (below)		UNINFECTED (10/50)	INFECTED (40/50)
Fecal production rate		0.0459 ± 0.039	0.0451 ± 0.037
Response to threat	Escape	3	19*
	Death feigning	7	21*
Response to light/dark	Neutral	4	15
	Shade	3	20
	Light	3	5
Weight		$0,25 \pm 0.042$	$0,22 \pm 0.041^{\dagger}$
Length		1.62 ± 0.11	$1.54 \pm 0.11^{\dagger}$
Abdomen width		0.53 ± 0.0028	$0.52 \pm 0.0049^{*}$
% of beetles with amputations		20% (2/10)	55%* (22/40)
Sex	Male	7	18
	Female	3	22

† = Significant difference ($P < 0.05$) by the Student's *t*-test; * = Significant difference ($P < 0.05$ or lower) by either the Kolmogorov-Smirnov or by one-way ANOVA (Pearson's r-correlation test).

Finally, in the analysis of influence of gregarine infection on female egg production, only 17 female beetles (out of 20) survived at the end of the experiment. The mean number of eggs laid by uninfected females (6/17) was 10.5 ± 7.96, whereas in infected insects (11/17) it was 8.72 ± 10.01 (such difference was non-significant by the Student's *t*-test). When the ratio "egg count/female weight" was employed for comparison (data not shown), the same non-significant result was obtained. Simple regression analysis showed no correlation at all between gregarine load and either egg count or egg count/beetle weight.

Scaurus punctatus as a Model System in Gregarine Studies

It is the first time that the beetle *S. punctatus* was the host used for gregarine studies, and a brief comment on its performance as a model system in the parasitology laboratory is necessary. Many adult beetles survived more than one year in captivity, reaching almost two years. Although beetles could have been maintained in an apple + Spam® diet, it was observed that they largely preferred decaying leaves and Spam® as food, ingesting apple in limited amounts, probably as a mere source of water. The addition of preserved pork meat is important to avoid cannibalism, as the protein content of leaves is limited. It was easy to maintain a infected population of beetles just feeding them with leaves from the capture site. There was no need of performing experimental infections in the laboratory. In fact, twelve beetles that were maintained at 24°C for eight months, and fed with naturally contaminated leaves) showed high gregarine prevalence. Nine (out of ten) surviving animals were infected, with a gregarine abundance (small phenotype) of 221.5 ± 169.24 protozoa per beetle, which is comparable to figures obtained in natural infections. On the other hand, the finding that *S. punctatus beetles* could be defaunated of gregarines by maintenance at 37°C for 9 days facilitated the performance of experimental infections in gregarine-free insects. In addition, the defaunated insects can remain uninfected if fed with sterile *P. hispanica* leaves, apple and preserved pork meat.

DISCUSSION

Epizootiological Study

The present study represents the first finding of *G. wharmani* in Spain and Europe. The total length observed in gregarine specimens found in *S. punctatus* gut or feces is in agreement with data originally published by Theodorides (1955c) related to *S. puncticollis* in Israel. The gregarine gamont shows a shallowly to broadly ovoid protomerite, clearly different of the obpanduriform shape usually observed in that of *G. ormierei*. In addition, *G. wharmani* deutomerite is also less cilyndrical in shape than that of *G. ormierei*. It seems likely that *G. wharmani* is present in *Scaurus* spp. throughout the Mediterranean basin, although in Spain its prevalence is very low, as we were unable to find it in a previous survey for gregarines

(Criado-Fornelio et al., 2017). On the other hand, the identification of the *Gregarina* sp n°2 is more difficult to confirm, due to the fact that the single gamont observed by us is smaller than the associated gamonts described by Theodorides (1955 a and b) in *Phylan planiusculus*. However, the morphology of the protomerite, along with the absence of any post-nuclear pigment stain suggests that the isolate from Alcalá de Henares does not belong to the species *G. maculata* described by Leger and Dubosq (1904). The prevalence of this species (*Gregarina* sp n°2) is low, and the scarce number of specimens observed by us and other authors in *Scaurus/Phylan* beetles intestine suggests that there might be another host species (to be discovered yet) best suited for the survival of this protozoa.

Implications of the Observed Variations of Epimerite Morphology in *G. ormierei*

According to Theodorides (1958a), *G. ormierei* shows an orbicular epimerite, and our previous observations in *S. punctatus* seemed to confirm such observation (Criado-Fornelio et al., 2017). However, we have pointed also that the morphology of *G. cavalierina* and *G. ormierei* were quite similar. The main difference between both species was the epimerite shape, that was conical-tubular (instead of orbicular) in the former (*G. cavalierina*), according to Filipponi (1952a) and Theodorides (1952). Unfortunately, in the original report on the erection of the species *G. cavalierina* by Blanchard (1905), the morphology of the trophozoite and its epimerite was undescribed. The current data available suggest that the epimerite of *G. ormierei* is initially spherical, but latter it evolves towards a tubule-like structure. Further proof of such feature requires additional histological sections examination of the infected gut cells in beetles, a study not done in the present work. If such a finding is confirmed, the species *G. ormierei* should be considerd invalid and it would remain as a synonym of *G. cavalierina*.

Biology of *G. ormierei* Infection

Previous studies by Theodorides (1955a, 1958a) did not report the complete life cycle of *G. ormierei*. Data presented here show similarities between the biology of this gregarine and that of *G. cavalierina* as reported by Blanchard (1905) and Filipponi (1952a, b and 1953). However, the duration of the life cycle remained unknown. In the present study, we observed that the life cycle of *G. ormierei* lasts around 10-25 days, depending on phenotype. These data are in the range described for other gregarines (Jhony and Whitman, 2005; Smith et al., 2007). In spite of the similar biology of *G. ormierei* and other Gregarinidae, some details on gametocyst dehiscence described in the present work are original and deserve further comment. It is a common belief that indehiscent gametocysts of gregarines either does not contain oocysts or contains non-viable oocysts. A viable gametocyst is that showing dehiscence (Clopton and Janovy, 1993). However, it has been demonstrated here that when apparently unripe gametocysts were included in beetle food and used for infecting *S. punctatus*, a noticeable proportion of these were viable. This may be due to the fact gametocysts are contained in a relatively resistant chitinous structure (at least in the small phenotype of *G. ormierei*). Thus, dehiscence and oocyst dispersal is obstaculised and probably both processes have become secondary for survival of small phenotype gregarines in *G. ormierei*. Although gametocyst inclusion in digestive membranes seems to be irrelevant for the big phenotype due to its bigger size, our results showed that dehiscence must be as well problematic since its dehiscence rate in air were quite low. Therefore, the ability of gametocysts to produce oocysts without dehiscence seems to be a crucial adaptation for a gregarine that parasitizes a beetle species adapted to xeric environments (Cartagena and Galante, 2003). Such biological feature, along with the gregarious behaviour of *S. punctatus* (López-Pérez, 2010), may help to explain the high prevalence found in our sampling area. Some authors suggested that a rapid superficial sterilization of gregarine gametocysts with 95% ethanol improves dehiscence rate (Filipponi, 1952a). Nevertheless, we did not assay such procedure, as the objective was to obtain results comparable to that occurring in nature.

Gregarina ormierei oocysts may show or not polar plates, which indicates that their presence or absence relies on whether their liberation occurs in chains or in masses. Such observation confirms that these structures play a mere adhesive role. They ensure the integrity of linear oocyst chains inside the tightly enclosing sporoduct membrane, a hypotesis previously suggested (but not addressed) by Ghose et al. (1986). There are some fascinating features related to *G. ormierei* oocysts, such as the presence of an excystation vacuole and the extrusion through an axial tubule. Such type of excystment is unusual in the Apicomplexa, but common in Ciliates (Müller, 2007). It seems that the process occurring in gregarines might be convergent with that occurring in Ciliates. The latter organisms show an active expansion of the excystment vacuole, caused by water intake in cysts (Funadani et al., 2013). Since the density of *G. ormierei* oocysts in extrusion process is lower than in resting oocysts, it is very likely that the water content actively increases in the resistant stage during excystation. Water intake might occur through the emergent blunt ends of the axial tubule. Nevertheless, ultrastructural studies by TEM are necessary to address such hypothesis, as mentioned above.

Defaunation of gregarines infecting *Tenebrio molitor* is possible keeping beetles at 37°C for 6 days (McDougall, 1942). In our experiments, the period needed to eliminate gregarines in *S. punctatus* was slightly longer. This may be due to the African origin of these beetles (Español, 1960); their gregarines may be better adapted to warm conditions. The finding that most small phenotype gregarines survived when beetles are maintained one month at 4°C is meaningful. Since *Scaurus* life span is of up to two years (Cartagena and Galante, 2003), such ability suggests that small gregarines are capable to survive winter periods in insects hibernating in soil. In contrast, big phenotype gregarines affected by a cold period were motionless and depleted of granules, probably with total loss of viability. Nevertheless the effect of moderately low temperatures in gregarines is not always detrimental, as pointed out in *G. cubensis* in cockroaches kept at 15°C (Smith et al., 2007). The big phenotype of *G. ormierei* seems better adapted to live at warmer temperatures, as judged by the experiments carried out at 32 and 37°C. However, it is sure that some other unknown factors influenced the development of the big form, since not all gregarines switched their phenotype in the experimental group kept at 32°C.

The frequency of gregarine infection may change throughout the year. In India, a high degree of humidity in the rainy season seemed to exacerbate gregarine prevalence (Ghose et al., 1986; Ghose and Haldar, 1989), as it favoured both gametocyst dehiscence and oocyst survival. In temperate regions in the USA, both temperature and humidity played essential roles in determining parasite abundance (Bunker et al., 2013). Seasonal variations in abundance of a single gregarine species occur, but the present work is the furst to report that phenotype expression varies throughout the year. It seems that in early/middle spring adult beetles infected by small phenotype gregarines are common, since this stage causes infection at relatively low temperatures in central Spain. Moreover, some of these gregarines may have survived the winter period in the intestine of beetles as well. Meanwhile, big phenotype gregarines are practically absent, due to their inability to survive the winter in the beetle gut or to cause infection at low temperatures. The abundance of the small phenotype decreases steadily in late spring, maybe due to its lower tolerance to the heat and lack of humidity, whereas the big phenotype becomes much more frequent. A peak in gregarine prevalence is evident in June, just after a peak in humidity (associated to rain) and a gradual increase in temperature. Indeed both factors positively affect insect activity and gregarine dissemination in diverse geographic locations (Baz-Ramos, 1986; Locklin and Vodopich, 2010). However, populations of both gregarine phenotypes decline when temperature reaches a maximum in July/August along with a minimum in rainfall for the same period. Such hostile environmental conditions negatively affect both beetles and protozoa. The present data on parasite population dynamics in *G. ormeirei* are in quite good agreement with the evolution of parasite abundance described for *Steganorhynchus dunwoodi* (a gregarine parasite of *Ischniura verticalis*) by Bunker et al., (2013). The short life cycle observed in the small phenotype seems to be an adaptation for the changing climatic conditions in Spain in spring, with alternance of hot and cold/rainy periods. Production of viable gametocysts in 10 days or less in this phenotype may favour a very rapid spread of gregarine infection under such circumstances.

Possible Detrimental Effects of *G. ormierei* Infection

Little is known about the pathogenicity of gregarines in different host species (Desportes and Schrevel, 2013). Some authors reported that gregarines caused no harm to their hosts and could be considered as commensals (Rodríguez et al., 2007), unlike findings shown in the present work. In our contribution, gregarines were abundant in insects of low weight and reduced size, which suggests a detrimental effect of *G. ormierei* infection in *S. punctatus* beetles. There are indeed examples of differences in gregarine prevalence between larvae of different sizes in coleopterans as *Tenebrio molitor* (Ruhnke and Janovy, 1990), but the effect found in that case was just the opposite: as larval size increased, an increase in prevalence of *G. cuneata* preceded a more substantial increase in parasite relative density. Unfortunately, these authors did not study the influence of gregarine load on the weight of beetle larvae. In agreement with our findings, possible detrimental effects of gregarines in insects (with evident loss of weigth as well) have been reported for coleopterans (Purrini and Kel, 1989), ortopterans (Harry, 1970 and Zuk, 1987) and odonates (Siva-Jothy and Plaistow, 1999).

It was intriguing to observe that gregarines were slightly more frequent in beetles with bold behaviour (that is, those not fleeing from threat) and with amputations, albeit both factors showed lesser influence on gregarine load than beetle weight and length. It can be speculated that incautious beetles may present greater mobility, suffering more encounters with predators/competitors and hence, more possibilities of limb or antenna amputation. Moreover, according to Krams et al. (2013) non-cautious insects show higher metabolic rates and ingest more food. Both features (greater mobility plus high food intake) make insects prone to harbour increased populations of intestinal gregarines, as observed as well in female odonates by Hecker et al., (2002) and Bunker et al., (2013).

CONCLUSION

Scaurus punctatus and *G. ormierei* in tandem are a suitable laboratory model in future gregarine studies, comparable to others previously established (Filipponi 1952a; Ruhnke and Janovy, 1990). Maintenance of insects (either infected or uninfected) in the laboratory was not problematic. *Scaurus punctatus* is relatively long-lived, which permitted to develop long-lasting experiments. Moreover, keeping gregarine-free or gregarine-infected beetles was relatively easy.

Finally, it must be underlined that *G. ormierei* is a parasite of paramount importance for coleopterans in the Old World, due to its extraordinary biological features: potential pathogenicity, high infection rate, relatively low parasitic specificity and wide geographic distribution. Moreover, taking into account the fact that *S. punctatus* is a very abundant species and that pesticides may cause effects on non-target organisms (such as detritivore insects, according to Souza et al., 2012), the presence of pathogens with high prevalence is particularly important. Indeed high parasite loads may exacerbate any negative effect caused by insecticides in beetles, which may lead to undesirable consequences in terrestrial ecosystems.

ACKNOWLEDGMENTS

We wish to thank Mr. Antonio Priego and Mr. José Antonio Pérez (Microscopy Unit - CAI Medicina y Biología de la Universidad de Alcalá). Mr Ángel Pueblas (Scientific Image and Photography Unit - CAI Medicina y Biología de la Universidad de Alcalá) also provided expert help with photographic work. The present work has received economical support by the Universidad de Alcalá - 3023 MF 100 department funding.

REFERENCES

Anonymous. (2008). Libros de caracterización agro-climática de las provincias españolas. Datos de la provincia de Madrid. *Publicaciones de la biblioteca del Ministerio de agricultura, alimentación y medio ambiente* [Book series on the agroclimatic characterization of the Spanish provinces. Data on the province of Madrid. *Publication Services of the Ministry of Agriculture, Food and Environment*]. Article available online:www.magrama.gob.es/ministerio/pags/Biblioteca /fondo/pdf/2866_9.pdf.

Baz-Ramos, A. (1986). Sobre la estacionalidad de las comunidades de mariposas (Lepidoptera) de la zona Centro de la Península Ibérica [Comments on the seasonality of ecological communities of butterflies (Lepidoptera) in the Central Iberian Peninsula]. *Bol Asoc Esp. Entom*, *10*, 139-157.

Blanchard, L. F. (1905). Deux gregarines nouvelles parasites des Tenebrionidés des Maures [Two new gregarines parasites of Tenebrionids in Marocco]. *C R Ass Fr Avan Sci*, *33*^{eme} *Session*, Grenoble, 1904, 923-928.

Bunker, B. E., Janovy, J., Tracey, E., Barnes, A., Duba, A., Shuman, M. & Logan, J. D. (2013). Macroparasite population dynamics among geographical localities and host life cycle stages: Eugregarines in *Ischniura verticalis*. *J Parasitol*, *99*, 403-409.

Cartagena, M. C. & Galante, E. (2003). Ecología de las especies de *Scaurus* Solier, 1836 en el sudeste ibérico (Coloeptera: Tenebrionidae) [Ecology of *Scaurus* spp. Solier, 1836 in Southwestern Spain]. *Ses Entomol ICHN-SCL*, *13*, 37-46.

Clopton, R. E. (2004). Standard nomenclature and metrics of plane shapes for use in gregarine taxonomy. *Comp Parasitol*, *71*, 130-140.

Clopton, R. E. (2009). Phylogenetic relationships, Evolution and systematic revision of the Septate gregarines (Apicomplexa: Eugregarinorida: Septatorina). *Comp Parasitol*, *76*, 167-190.

Clopton, R. E., Percival, T. J. & Janovy, J. (1991). *Gregarina niphandroides* (Apicomplexa: Eugregarinida) described from adult *Tenebrio molitor* (Linneus) with oocyst descriptions of other gregarine parasites of the yellow mealworm. *J Protozool*, *38*, 472-479.

Clopton, R. E., Percival, T. J. & Janovy, J. (1992). *Gregarina coronata* n. sp. (Apicomplexa: Eugregarinida) described from adults of the southern corn rootworm, *Diabrotica undecimpunctata howardi* (Coleoptera: Chrysomelidae). *J Protozool*, *39*, 417–420.

Clopton, R. E. & Janovy, J. (1993). Developmental niche structure in the gregarine assemblage parasitizing *Tenebrio molitor*. *J Parasitol*, *79*, 701-709.

Criado-Fornelio, A., Pérez-Serrano, J., Heredero-Bermejo, I. & Verdú-Expósito, C. (2013). New advances on the biology of gregarines (Protozoa: Apicomplexa): first finding of Eugregarinorida in terrestrial isopods in Spain by molecular methods. In: *Protozoa, Biology, Classification and role in disease*. Editors: C. Castillo and R. Harris, R.; Nova Science Publishers, Hauppage, NY, USA, pp. 109-130.

Criado-Fornelio, A., Verdú-Expósito, C., Martin-Pérez, T., Heredero-Bermejo, I., Pérez-Serrano, J., Guardia-Valle, L. & Panisello-Panisello, M. (2017). A survey for gregarines (Protozoa: Apicomplexa) in arthropods in Spain. *Parasitol Res*, *116*, 99-110.

Desportes, I. & Schrevel, J. (2013). *Treatise on Zoology - anatomy, taxonomy, biology. The gregarines, the early branching Apicomplexa.* First ed. Brill, Leiden, The Netherlands.

Español, F. (1960). Los *Scaurus* de España (Coleoptera, Tenebrionidae) [The *Scaurus* beetles of Spain]. *Eos*, *36*, 141-155.

Filipponi, A. (1952a). *Protomagalhaensia marottai* n. sp. (Gregarinidae) parassita di *Scaurus striatus*. *Riv Parassitol*, *13*, 143-156.

Filipponi, A. (1952b). Accrescimento relativo in due fenotipi di *Protomagalhaensia marottai* Filipponi 1952 [Relative growth of two fenotypes of *Protomagalhaensia marottai* Filiponni 1952]. *Riv Parassitol*, *13*, 217-234.

Filipponi, A. (1953). Sul grado di stabilita nei caratteri di *Protomagalhaensia marottai* (Sporozoa, Gregarinidae) [Comments on trait stability of *Protomagalhaensia marottai* (Sporozoa, Gregarinidae)]. *Riv Parassitol*, *14*, 137-163.

Funadani, R., Suetomo, Y. & Matsuoka, T. (2013). Emergence of the terrestrial ciliate *Colpoda cucullus* from a resting cyst: rupture of the cyst wall by active expansion of an excystment vacuole. *Microbes Environ*, *28*, 149-52.

Ghose, S., Sengupta, T. & Haldar, D. P. (1986). Two new septate gregarines (Apicomplexa: Sporozoea), *Gregarina basiconstrictonea* sp. n. and *Hirmocystis oxeata* sp.n. from *Tribolium castaneum* (Herbst). *Acta Protozool*, *25*, 93-108.

Ghose, S. & Haldar, D. P. (1989). Role of environmental factors in the incidence of two new species of apicomplexan parasites, *Hirmocystis lophocateri* sp.n. and *Hirmocystis triboli* sp. n. from coleopteran insects. *Acta Protozool*, *28*, 49-60.

Greenslade, P. J. M. (1964). Pitfall trapping as a method for studying populations of Carabidae (Coleoptera). *J Anim Ecol*, *33*, 301-310.

Harry, O. G. (1970). Gregarines: their effect on the growth of the desert locust (*Schistocerca gregaria*). *Nature*, *225*, 964 - 966.

Hecker, K. R., Forbes, M. R. & Leonard, N. J. (2002). Parasitism of *Enallagma boreale* damselflies by gregarines: sex biases and relations to adult survivorship. *Can J Zool*, *80*, 162–168.

Janovy, J., Detwiler, J., Schwank, S., Bolek, M. G., Knipes, A. K. & Langford, G. J. (2007). New and emended descriptions of gregarines from flour beetles (*Tribolium* spp. and *Palorus subdepressus*: Coleoptera, Tenebrionidae). *J Parasitol*, *93*, 1155-1170.

Krams, I., Kivleniece, I., Kuusik, A., Krama, T., Raivo, M., Rantala, M. J., Znotin, S., Freeberg, T. M. & Mand, M. (2013). Predation promotes survival of beetles with lower resting metabolic rates. *Entomol Experim Applic*, *148*, 94-104.

Leger, L. & Dubosq, O. (1904). Nouvelles recherches sur les gregarines et l'epithelium intestinal des Tracheates [New investigations concerning the gregarines of the intestinal epithelium of Tracheates]. *Arch f Protist*, *4*, 335-83.

Levine, N. D. (1971). Uniform terminology for the Protozoan Phylum Apicomplexa. *J Protozool*, *18*, 352-355.

Locklin, J. L. & Vodopich, D. S. (2010). Patterns of gregarine parasitism in dragonflies: host, habitat and seasonality. *Parasitol Res*, *107*, 75-85.

López-Pérez, J. J. (2010). Corología de los Scaurini Billberg, 1820 (Coleoptera, Tenebrionidae, Tenebrioninae) provincia de Huelva, Sur-Oeste de la Península Ibérica [Chorology of the Scaurini Billberg, 1820

(Coleoptera, Tenebrionidae, Tenebrioninae) in the province of Huelva, Southwestern Spain]. *Boln Assoc Esp Ent, 34,* 7-14.

McDougall, M. M. (1942). A study of temperature effects on gregarines of *Tenebrio molitor* larvae. *J Parasitol, 28,* 233-240.

Moreno, G. & Manjón, J. L. (2010). *Guía de Hongos de la Península Ibérica* [*A guide for the Funguses of the Iberian Peninsula*]. G. Moreno and J.L. Manjon Eds., Editorial Omega, Barcelona (Spain), pp. 1440.

Müller, H. (2007). Live observation of excystment in the spirotrich ciliate *Messeres corlissi. Eur J Protist, 43,* 95-100.

Purrini, K. & Kel, H. (1989). *Ascogregarina bostrichidorum* n. sp. (Lecudinidae: Eugegarinida), a new gregarine parasitizing the larger grain borer, *Prostephanus truncatus* Horn (1878) (Bostrichidae: Coleoptera). *Arch fur Protistenk, 137,* 65-171.

Rodríguez, Y., Omoto, C. & Gomulkievicz, R. (2007). Individual and population effects of Eugregarine, *Gregarina niphandroides* (Eugragrinida: Euigregarinidae), on *Tenebrio molitor* (Coleoptera: Tenebrionidae). *Environ Entomol, 36,* 689-693.

Ruhnke, T. R. & Janovy, J. (1990). Life history differences between two species of *Gregarina* in *Tenebrio molitor* larvae. *J Parasitol, 76,* 419-22.

Smith, A. J., Cook, T. J. & Lutterschmidt, W. I. (2007). Effects of temperature on the development of *Gregarina cubensis* (Apicomplexa: Eugregarinida) parasitizing *Blaberus discoidalis* (Blattaria: Blaberidae). *J Parasitol, 93,* 583-8.

Siva-Jothy, M. T. & Plaistow, S. J. (1999). A fitness cost of eugregarine parasitism in a damselfly. *Ecol Entomol, 24,* 465–470.

Souza, C. R., Sarmento, R., Venzon, M., Barros E., Rodrigues dos Santos, G. & Chaves, C. C. (2012). Impact of insecticides on non-target arthropods in watermelon crop. *Semin Cienc Agrarias, 33,* 1789-1802.

Theodorides, J. (1952). Inexistence du genre *Protomagalhaensia* Pinto (Sporozoa, Gregarinidae). Identité de *P. marottai* Filipponi avec *Gregarina cavalierina* Blanchard [The nonexistence of the genus *Protomagalhaensia* Pinto (Sporozoea, Gregarinidae). Synonymia of P.

marottai Filipponi with *Gregarina cavalierina* Blanchard]. *Riv Parassitol*, *13*, 211-216.

Theodorides, J. (1955a). Les eugregarines du genre *Gregarina* parasites des coleopteres Tenebrionidés [The eugregarines of the genus *Gregarina* parasites of Tenebrionid coleoptera]. *Ann Parasitol Hum Comp*, *30*, 5-21.

Theodorides, J. (1955b). Gregarines de coleopteres du Maroc [The gregarines of coleopterans in Marocco]. *Arch. Inst. Pasteur Maroc*, *V*, 3-14.

Theodorides, J. (1955c). Gregarines parasites des coleopteres tenebrionidés d'Israel. *Ann Parasitol Hum Comp*, *30*, 161-173

Theodorides, J. (1958a). *Gregarina ormierei* Theodorides 1955 (Eugregarina, Gregarinidae) retrouvée chez un Tenebrionidé de Turquie [The finding of *Gregarina ormierei* Theodorides 1955 in a Tenebrionid beetle in Turkey]. *Vie et Milieu*, *9*, 125.

Theodorides, J. (1958b). Deux nouvelles eugregarines parasites des *Hegeter* (Coleoptera: Tenebrionidae) des iles Canaries: *Gregarina joliveti* n.sp. (Stylocephalidae) [Two new eugregarines parasites of *Hegeter* (Coleoptera, Tenbrionidae) in the Canary Islands: *Gregarina joliveti* n.sp.]. *Ann Sci Nat Zool Biol Anim*, *20*, 105-110.

Theodorides, J. (1960). Quelques parasites de Coléoptères Ténébrionides d'Espagne [Parasites found in Tenebrionid coleopterans in Spain]. *Ann Parasitol Hum Comp*, *35*, 762-763.

Theodorides, J., Desportes, I. & Jolivet, P. (1964). Gregarines parasites de Coleopteres d'Ethiopie [Parasitic gregarines of Coleoptera in Ethiopia]. *Ann Parasit Hum Comp*, *39*, 1-31.

Theodorides, J., Desportes, I. & Jolivet, P. (1965). Gregarines parasites de Coleopteres Tenebrionidés de la region de Khartoum (Republique de Soudan) [Parasitic gregarines of Coleoptera in the Khartoum region]. *Bull Inst Français de l'Afrique Noire*, *27*, 139-164.

Theodorides, J., Desportes, I. & Jolivet, P. (1976). Gregarines de la Corée du Sud [Parasitic gregarines of South Korea]. *Ann Parasit Hum Comp*, *51*, 161-173.

Zuk, M. (1987). The effects of gregarine parasites on longevity, weight loss, fecundity and developmental time in the field crickets *Gryllus veletis* and *G. pennsylvanicus*. *Ecol Entomol, 12*, 349–354.

Revised by Dr. Amelia Buling (Former Professor of Biology at the Foundation University, Dumaguete City, The Philippines).

INDEX

A

Achelata (achelate lobster), 79, 83, 87, 100, 106
adaptation(s), 5, 12, 29, 33, 34, 65, 100, 113, 147, 149
adaptive radiation, viii, 2, 3, 32, 33, 57
adults, 65, 67, 106, 152
age, 13, 14, 42, 47, 52, 56
Anomura (anomuran), 79, 91, 112
ANOVA, 78, 85, 86, 97, 143
Aristeidae, 73
arthropod communities, 2, 3, 5, 6, 7, 8, 10, 12, 16, 18, 19, 20, 21, 26, 28, 29, 34, 40, 42, 48, 59, 60
arthropods, 3, 4, 5, 7, 8, 9, 11, 12, 13, 14, 15, 16, 17, 18, 19, 20, 21, 22, 24, 26, 27, 29, 30, 31, 32, 34, 37, 40, 51, 59, 60, 63, 103, 104, 116, 122, 123, 152, 155
associational susceptibility, 3, 4, 61
Astacidea (clawed lobster), 65, 66, 70, 79, 82, 88, 89, 91, 118
Axiidae, 79, 92

B

beetles, vii, ix, x, 19, 21, 35, 41, 46, 122, 123, 124, 125, 127, 128, 129, 130, 137, 139, 141, 142, 143, 144, 145, 146, 148, 149, 150, 151, 153, 154
Benthesicymidae, 73, 96
biodiversity, 7, 14, 17, 20, 22, 29, 38, 43, 44, 46, 47, 50, 51, 56, 59, 61
biology, vi, vii, ix, x, 17, 35, 36, 37, 42, 44, 45, 50, 56, 58, 61, 110, 111, 114, 116, 117, 118, 121, 122, 123, 124, 133, 142, 146, 152, 153, 156
biomass, 5, 6, 7, 12, 13, 16, 19, 24, 33, 39
biotic, 6, 19, 34, 53, 106
birds, 16, 51
Birgus latro, 79, 91, 112
Brachyura (crab), 79, 87, 92, 103, 106, 111, 112, 113, 115

C

canopy arthropods, 59
Carcinus maenas, 79, 87, 88, 92, 103, 106, 113, 115
Caribbean, 112, 119

D

E